THE
PRACTICAL
PYROMANIAC

THE
PRACTICAL
PYROMANIAC

Build Fire Tornadoes, One-Candlepower Engines,
Great Balls of Fire, and More Incendiary Devices

WILLIAM GURSTELLE

CHICAGO
REVIEW
PRESS

Library of Congress Cataloging-in-Publication Data
Gurstelle, William.
  The practical pyromaniac : build fire tornadoes, one-candlepower engines,
great balls of fire, and more incendiary devices / William Gurstelle. — 1st ed.
      p. cm.
  Includes bibliographical references and index.
  ISBN 978-1-56976-710-8 (pbk.)
  1. Fire. 2. Combustion, Theory of. I. Title.

TP265.G87 2011
541'.361—dc22

                                                                      2010053910

The author and the publisher of this book disclaim all liability incurred in
connection with the use of the information contained in this book.

Questions? Comments? Visit www.ThePracticalPyromaniac.com
Follow the Practical Pyromaniac:
www.facebook.com/PracPyro
www.twitter.com/PracPyro

Cover design: John Yates at Stealworks.com
Cover photo: David Howells
Interior design: Scott Rattray
Interior illustrations: Todd Petersen

Copyright 2011 by William Gurstelle
All rights reserved
First edition
Published by Chicago Review Press, Incorporated
814 North Franklin Street
Chicago, Illinois 60610
ISBN 978-1-56976-710-8
Printed in the United States of America
1 2 3 4 5

Dedicatory Clerihew

The Hansen family
Is known for their hospitality,
Without being specific,
In all ways they're terrific.

Contents

Introduction   ix

1   Keeping Safety in Mind   1

2   The Flame Tube   7

3   The First Lights   31

4   The One-Candlepower Engine   47

5   The Fire Drill   55

6   The Burning Ring of Fire   69

7   The Hydrogen Generator and
    the Oxygenizer   81

8   Exploding Bubbles   101

9   The Fire Piston   113

10 The Arc Light   123

11 Fireproof Cloth and Cold Fire   131

12 The Extincteur   147

13 The Photometer   155

14 Thermocouples   165

15 Technicolor Flames   175

16 The Fire Tornado   183

17 Great Balls of Fire   193

Epilogue   201
Bibliography   205
Index   207

Introduction

THE PARADOX OF FIRE

Fire is the most important agent of change on earth. It makes our cars and airplanes move, it purifies metals, it cooks our food. It also destroys forests and pollutes the atmosphere. Fire is also one of the most paradoxical forces in nature. Sometimes it's incredibly difficult to light a much-desired campfire and keep it going, while at other times unwanted fires start far too easily.

To Greek philosophers of the Classical era, fire was a tangible, material thing. The legends they repeated held that noble Prometheus purloined fire from Mount Olympus and secretly gave it to human beings, much to the chagrin of an angry Zeus.

As Greek civilization progressed, legends became insufficient; people sought to understand fire on a more scientific basis. The first major nonmythological theorist was the Greek scholar Empedocles, who devised the earliest well-known explanation of the nature of the world. Everything, he said, was made up of four elements: earth, air, water, and fire. This was called the Four Element hypothesis. Aristotle refined that a bit, and for the next 2,000 years it was accepted with only minor modifications as the cosmological basis for the entire universe.

The hypothesis stated that everything in the world is composed of these four elements; the only difference between all the things we see or touch is the relative abundance of the four constituent components. Wood, according to Aristotle, is a composition of earth and fire, as evidenced by the way wood burns. Nonburning rock is mostly earth, with perhaps a bit of water added.

In the Middle Ages, this explanation no longer fit the results of many experiments that involved fire. Many alchemists had excellent experimental technique and analyzed a number of chemical processes in their quest to turn base metals into gold. But when fire was involved, their experiments yielded results that didn't jibe with the classical Four Element worldview. The pillars of cosmological doctrine were crumbling away. The alchemists were beginning to suspect that the world was more complex than they had been taught.

In the 1700s, a new way of thinking called phlogiston (flo-jis'-ton) theory came into fashion. This theory, which was promoted by many of the leading scholars of the age, held that fire was caused by the liberation of an undetectable chemical substance—phlogiston—which was bound up inside all things that could be made to burn. Those items that possessed phlogiston ignited and combusted; those without it did not. Phlogiston theory made sense for a while. But like the Greek notion of matter, phlogiston was merely an expedient, a theory cobbled together to describe the things that even close observations of the world could not otherwise explain.

At the end of the 18th century, improvements in experimental technique combined with the astute observations of a new generation of enlightened thinkers led to a much better understanding of the world in general and science in particular. It started with important discoveries by Benjamin Thompson, Count Rumford; Joseph Priestley; and Henry Cavendish. The work they did laying the foundations of modern chemistry was built upon by others, notably John Dalton and Antoine Lavoisier, until a new and correct interpretation of the phenomenon of fire emerged. At the turn of the 19th century, scientists were finally beginning to truly comprehend fire.

At that time there was a lively, collegial, and incredibly productive community of scientists fascinated by fire. During a fairly short window of time, a few years on either side of the year 1800 and centered around the Royal Institution in London, a surprisingly small but interconnected community of scientists solved the mysteries and banished the superstitions surrounding fire, finally allowing scientific understanding to take hold.

It was not a direct path. There was plenty of meandering and zigzagging through half-correct theoretical deductions and unexpected laboratory results. But eventually, a body of true and practical knowledge was accumulated. It was these turn-of-the-19th-century "natural philosophers" (the word "scientist" was not coined until the 1830s) who paved the way for modern scientists to understand the true nature of fire.

Isaac Watts was one of the key contributors to the advancement of pre–Industrial Revolution science, but he wasn't a scientist. He was a 17th-century English logician and musician, best known as a composer of Christian hymns. (His most famous work is "Joy to the World.") More than that, he was an important theorist on the nature of learning, a father of scientific and logical pedagogy. His influence on the scientists and experimenters who appear in these pages was immense.

Watts shared his philosophy on understanding the world in several highly regarded books, his most famous being *The Improvement of the Mind*. Written in 1815 toward the end of his life, it had tremendous influence on generations of students and teachers. It is still in print and popular even now, 200 years after Watts wrote it.

Watts's books were standard issue to generations of Oxford and Cambridge University students. His ideas served as one of the foundations for learning logical thought, shaping European society for more than a hundred years. Many suggestions for bettering oneself flow through the pages of Watts's books. Foremost among them, Watts urged his readers to improve their minds in five different ways, which he called his "five pillars of learning." Through the technique of the five pillars, Watts hoped to improve the lot of the world.

> There are five eminent means or methods whereby the mind is improved in the knowledge of things, and these are: observation, reading, instruction by lectures, conversation, and meditation.

All of these methods, wrote Watts in the pedantic, pointed, yet elucidating style of 18th-century English self-help authors, are important and useful in improving the mind. But of all the methods, judged Watts, observation is the foundation upon which all other learning methods rest. As Watts explains, "We may justly conclude, that he that spends all his time in hearing lectures or pouring upon books without observation . . . will have but a mere historical knowledge of learning, and be able only to tell what others have known or said on the subject."

Reading a book like *The Practical Pyromaniac* is one of the five Wattsian ways in which knowledge can be acquired. What sets this book apart from others, however, is the integration of all of the methods Watts recommends, including lecture and personal observation.

Besides providing the stories of great scientists, experimenters, and practical geniuses—those who played with fire and in so doing came up with new and important ideas and inventions—*The Practical Pyromaniac* contains numerous peripatetic projects and experiments. Further, there are video demonstrations on the Internet designed specifically to integrate with the information.

While it is not necessary to view the videos, undertake the experiments, or read the other books listed in the bibliography, if you do attempt a few of the projects and view some of the video lectures and demonstrations developed in conjunction with this book, your experience will be enhanced. The Internet addresses of the relevant videos are shown in boxes that look like this.

---

www.video1.ThePracticalPyromaniac.com

⚠ 1 ⚠

KEEPING SAFETY IN MIND

We've always been told to be careful around fire for good reason: fire can be dangerous! So, you may be wondering, "Doesn't this book advocate playing with fire?"

Indeed it does, but this type of play has both purpose and reason. The projects have been designed to be exciting and fun, and, very often, the result of a successfully rendered experiment is something catching on fire or an explosion. But there is a method to this madness, and what's described here is much different from foolishly playing with matches or causing destructive fires.

It's very important to note that if you don't follow the directions closely many of the projects in *The Practical Pyromaniac* could harm you or people around you. Always follow the project instructions closely. Don't make changes to the materials or construction techniques. This can lead to unexpected and unintended results. Further, be aware that some projects describe using materials in ways and under conditions that depart from the manufacturers' recommendations, so leaks, breaks, and other failures may be more likely to occur.

A VERY IMPORTANT MESSAGE

The projects described in the following pages have been designed with safety foremost in mind. However, as you try them out, there is still a possibility that something unexpected may occur. It is important that you understand that neither the author, the publisher, nor the bookseller can or will guarantee your safety. When you try the projects described here, you do so *at your own risk*.

Some of the projects have been popular for many years, while others are new and were designed specifically for this book. Unfortunately, even if you do everything correctly, something could still go wrong and cause damage to both property and people. The likelihood of such an occurrence is remote as long as you follow the directions, but things can still go wrong. Always use good common sense and remember that all experiments and projects are carried out at your own risk.

Be aware that each city, town, and municipality has its own rules and regulations, some of which may apply to projects described in *The Practical Pyromaniac*. Further, local authorities have wide latitude to interpret the law. Therefore, you should take time to understand the rules, regulations, and laws of the area in which you plan to carry out these projects. A check with local law enforcement can tell you whether the project is suitable for your area. If not, there are plenty of other places where all of the projects here can be undertaken safely and legally. If in doubt, be sure to check first!

GENERAL SAFETY RULES

The following are important general safety rules. Note that each chapter also provides specific safety instructions.

1. The experiments described here run the gamut from simple to somewhat complex. Some experiments involve the use of fire, volatile materials, and chemical reagents. The projects described here are designed for adults or, at a minimum, to be closely supervised by adults. Adult supervision is mandatory for all experiments and projects.

2. Read the entire project description carefully before beginning the experiment. Make sure you understand what the experiment is about and what you are trying to accomplish. If something is unclear, reread the directions until you fully comprehend the entire experiment.

3. Don't make substitutions for the chemical powders or liquids indicated for use in each experiment. Specificity counts. For example, substituting methanol for ethanol or a chlorate for a chloride could have serious consequences.

4. Use only the quantities of chemicals listed in the project instructions. Don't use more than specified.

5. Don't substitute materials or alter construction techniques. Don't take shortcuts.

6. Read and obey all product label directions, including the material safety data sheets that accompany chemicals. Acids and other chemicals must be handled appropriately as described on the container labels. Manufacturers are most familiar with how their products work, and their advice takes precedence.

7. Prior to performing any project, remove and safely store all cans or bottles containing flammable substances. Maintain a hazard-free radius of at least 50 feet around the area in which you plan to work.

8. Wear protective eyewear when indicated in the directions. Some experiments call for gloves, proper ventilation, and so forth. Always follow the safety guidelines given in the directions.

9. Visit www.ThePracticalPyromaniac.com prior to performing any project or experiment to check for safety updates and other important information.

10. Keep an all-purpose fire extinguisher ready and close by. All-purpose (often labeled A-B-C) fire extinguishers work on all types of fires. Choose a dry chemical extinguisher that is rated for multiple-purpose use. Widely available and inexpensive, dry chemical extinguishers contain a powdery extinguishing chemical and use a compressed, nonflammable gas as a propellant.

How to Use a Fire Extinguisher

Here's an easy acronym for fire extinguisher use:

P A S S   Pull, Aim, Squeeze, and Sweep

PULL   the pin at the top of the extinguisher that keeps the handle from being accidentally pressed.

AIM   the nozzle toward the base of the fire.

STAND   approximately 8 feet from the fire if you can and SQUEEZE the handle to discharge the extinguisher.

SWEEP   the nozzle back and forth at the base of the fire.

Remember this: The instructions and information provided here are for your use without any guarantee of safety. Each project has been extensively tested in a variety of conditions. But variations, mistakes, and unforeseen circumstances can and do occur. Therefore, all projects and experiments are performed at your own risk! If you don't agree with this, then do not attempt any experiments or projects.

Finally, there is no substitute for common sense. If something doesn't seem right, stop and review what you are doing. You must take responsibility for your own safety and the safety of others around you.

OBTAINING SUPPLIES AND EQUIPMENT

The projects in *The Practical Pyromaniac* have been designed to utilize common, inexpensive, and easy-to-find materials whenever possible.

The materials and tools required for most projects can be purchased at retail establishments such as hardware stores, craft stores, artist supply stores, and lumberyards. Harder-to-find parts such as polycarbonate plastic or nichrome wire may be sourced from the large online retailer McMaster-Carr (www.mcmaster.com). In addition, a large number of mail-order and online retailers, such as the following, sell chemicals and laboratory equipment that may be difficult to source locally.

United Nuclear
PO Box 851
Sandia Park, NM 87047
www.unitednuclear.com

A wide variety of chemicals and laboratory supplies including many unusual items others often don't stock

The Science Company
95 Lincoln Street
Denver, CO 80203
www.sciencecompany.com

Extensive selection of glassware, lab burners, chemicals, and safety equipment carried in a variety of sizes and quantities

Home Science Tools
665 Carbon Street
Billings, MT 59102
www.homesciencetools.com

Caters to homeschoolers and provides a good selection of laboratory equipment and chemicals in reasonable quantities

In addition, links to sources of materials are provided for many projects. Look for:

www.Project1.ThePracticalPyromaniac.com

Some items, such as pipes, pipe fittings, or lumber are too big to be shipped economically. They are commonly available locally at hardware, drug, or grocery stores.

A NOTE ABOUT UNITS

In most project descriptions, this book uses the American system of measurement of inches, feet, ounces, pounds, and so forth. However, in projects that are chemical in nature, such as the Oxygenizer and Exploding Bubbles, metric units are used. This is because most chemists use metric measurements and laboratory gear is scaled and graduated metrically.

A number of measurement converters are available for free on the Internet, including www.onlineconversion.com and www.convertit.com.

⚠ 2 ⚠

THE FLAME TUBE

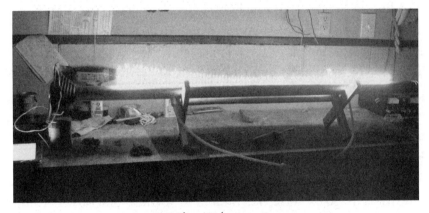

2.1 Flame Tube in action

I propose to bring before you, in the course of these lectures, the Chemical History of a Candle. There is no better, there is no more open door by which you can enter into the study of science than by considering the physical phenomena of a candle. There is not a law under which any part of this universe is governed which does not come into play and is not touched upon by these phenomena.

—Michael Faraday, 1860, from
the lecture "The Chemical
History of a Candle"

Soon, we'll meet many of the great scientists of the past who first explored the nature of fire. Besides being intellectual giants, they had something else in common. These fellows who furthered our knowledge of the chemistry and physics of fire were, generally speaking, grumpy old men. John Dalton was widely known to be "stiff" and "gruff," Antoine Lavoisier could be authoritarian, Benjamin Thompson dyspeptic, George Manby an angry megalomaniac, and Henry Cavendish . . . well, Cavendish was Cavendish.

If you were to go backward in time and meet them face to face, chances are you'd find the experience off-putting. It's a bit ironic that many of the men who explored heat and fire were a bit cold themselves.

On the other hand, if you could meet Michael Faraday, you'd find him warm and approachable. Faraday was easygoing, pleasant, and sociable. More importantly, he was perhaps the finest experimental scientist in the history of chemistry and physics. Faraday was a brilliant polymath responsible for a huge number of outstanding advances in the theory and application of magnetism, electricity, and chemistry. Through self-education, perseverance, and his own hard work, Michael Faraday, the son of an English blacksmith, became one of the most honored scientists in history.

Although he could have made a handsome salary working for industry, his goals were not fame and fortune but rather a hope that his discoveries would lead to a better life for all. Faraday spent nearly his entire life inhabiting a relatively small suite of rooms on the upper floors of the Royal Institution in London, where he worked in the facility's fine laboratory for a meager salary as director of its chemistry laboratory. But this arrangement suited him, as it allowed him to concentrate without distraction on his projects and experiments. His biographers wrote that Faraday "loved the labor far more than the wage."

Faraday employed splendid experimental technique, carefully applying the knowledge he learned from one experiment to the next. In so doing, he laid the foundation for an astonishing amount of the technology we use daily, including electrical power generation and many types of electrochemistry. Beyond that, he made several important discoveries in chemistry, metallurgy, and optics. It might not be too big an overstatement to declare that every time we flip a switch and a light goes on, the roots of the technology can be traced to Faraday's experiments in the laboratory at the Royal Institution.

A considerate and unpretentious man, Faraday did more than just labor in his laboratory. He found time for his wife and his church as well as for a few simple pleasures, such as taking nieces and nephews to the zoo or playing games or taking long walks. And he loved to introduce others, young and old alike, to the world of science in which he found such happiness.

— — — — — — — — — — —

By 1826, Faraday was well established in Europe's scientific community and had become keenly interested in teaching and encouraging young people toward scientific careers. He began a tradition in science education that continues to this day. Every December, a world-class scientist addresses a packed crowd made up mainly of young people at the Royal Institution in London's Mayfair neighborhood. The purpose is to provide insight and inspiration to the next generation of scientists and engineers. The Christmas Lectures, as they are called, are among the most important and well-known scientific discourses on the planet.

Faraday hoped his lectures would encourage his listeners to share his interest and awe in all things scientific and natural. Using as many of Isaac Watts's pillars of learning as possible, he hoped to inspire his audiences to move beyond reading and listening to personal observation and perhaps even to experimentation.

Under the scrutiny and analysis of Faraday's sparkling intellect, even everyday, commonplace phenomena and events, such as the flame of a candle or the swinging of a pendulum, became intriguing and worthy of deeper understanding and further investigation.

Let us now consider how wonderfully we stand upon the world. Here it is we are born, bred, and live, and yet we view these things with an almost entire absence of wonder to ourselves respecting the way in which all this happens.

[Except for a] few enquiring minds, who have ascertained the very beautiful laws and conditions by which we do live and stand upon this earth, we should hardly be aware that there was anything wonderful in it.

—The beginning of Faraday's 1860
Christmas Lecture to young
people at the Royal Institution

In 1860, Faraday ascended to the raised podium at the Royal Institution to give his final series of lectures there, this time entitled "The Chemical History of a Candle." As he lectured, Faraday animated his lessons with demonstrations, taking time to make his points logical and obvious to the assembled young people. By all accounts, Faraday was an enthralling speaker, neither talking down to his audience nor being in the least obtuse or obscure. He was, wrote an auditor, "the complete master of the situation. He had an irresistible eloquence, which compelled attention and insisted upon sympathy. . . . A pleasant vein of humor accompanied his ardent imagination. His experiments [were] true illustration for his arguments."

Sitting halfway back in the Royal Institution's lecture hall on this occasion was William Crookes. Crookes was a preeminent science journalist of his day and would go on to become a world-class scientist in his own right. He discovered the element thallium and invented many scientific instruments, including a still-popular science toy called a Crookes Radiometer, a spinning armature in a glass bulb that is powered only by sunlight.

In 1860, the 28-year-old Crookes, with his long, pulled mustache and full beard, must have looked out of place among the children and teenagers in the audience. But he was there for a purpose. Crookes's goal in attending this science lecture was to write down every word that Faraday spoke and make sketches of the experiments he performed. Crookes well understood the importance of Faraday's lectures and wanted to save them for posterity.

How lucky for us that he did so. The six lectures that make up "The Chemical History of a Candle" series are among the clearest and most straightforward explanations of fire ever provided. Since 1860 and continuing to this day, scores of editions of Crookes's transcript of that important moment have provided illumination, motivation, and recreation for generations of nascent scientists and researchers.

But that's not to say that the information contained in Crookes's transcript is ideal for modern readers. The manner in which science is taught has changed considerably in the 150 years since Faraday. While the information in the lectures is still vital and important, there have been some refinements in pedagogy as well as considerable progress in instrumentation and the way experiments are carried out.

2.2 Faraday lecturing

This section begins with updated versions of Faraday's classic demonstrations and projects and concludes with the exciting project known as the Rubens Flame Tube. The updates to the original Faraday projects are true to their scientific roots and in strict accord with the goal of his original lectures, but they differ from the originals in both presentation and scope.

The first purpose of updating the lectures is to present Faraday's brilliant experiments in a manner with which contemporary experimenters and students are familiar, using the modern inexpensive instruments and laboratory supplies available to us that were not available to Faraday. The second purpose relevant to his book is to focus on the information that helps us understand the true and basic nature of the phenomenon of fire. Faraday's Christmas Lectures dealt extensively with understanding the nature of fire, but he went far beyond that, explaining such topics as the chemical composition of the atmosphere and even the biology of animal respiration.

The original lecture series "The Chemical History of a Candle" consisted of six lectures, but it is the material from the first two, "A Candle: Sources of Flame" and "A Candle: Brightness of the Flame," that is used here.

– – – – – – – – – –

And now, I have to ask your attention to the means by which we are enabled to ascertain what happens in any particular part of the flame—why it happens, what it does in happening, and where, after all, the whole candle goes to. As you know, a candle brought before us and burned, disappears—and this is very curious.

Always remember that when a result happens, you should say, "What is the cause? Why did it occur?" and in the course of time you will find the reason.

— Michael Faraday's opening
remarks at his second 1860
Christmas Lecture

By methodically answering a few one-word questions—why, what, and how—Faraday was able to explain the nature of fire to his audience. Being a great experimenter, Faraday devised simple projects and experiments to test individual ideas and weave the answers together into a complete understanding of the issue. That was his method of coming up with a clear picture of complex phenomena, and an effective method it was and still is. Like Faraday, we start small and attempt, through experimentation and deduction, to understand the nature of fire by examining the physical processes occurring within the flame of a candle. Once a basic understanding of fire is obtained, we can build on what we've learned to undertake some pretty incredible projects.

KEEPING SAFETY IN MIND

1. These experiments are for adults or for those under the close supervision of an adult.
2. All experiments and projects should be performed on a heat-proof work surface. Cover wood or plastic surfaces to protect them from damage.
3. Have a bucket of water or fire extinguisher at hand.

- - - - - - - - - - -

Shaping a Candle Flame

"Air is absolutely necessary for combustion," wrote Faraday. That air is necessary for a continuing flame is obvious if you notice how the shape of a candle flame is affected by air flowing near it. When the air flow changes, so does the shape of the flame. This simple experiment in flame shaping proves this important point.

www.Project1.ThePracticalPyromaniac.com

Materials

- Matches
- New candle

1. Light the candle and look carefully at the area just below the wick. As the wax liquefies, it is transported to the flame via the wick. The candle's top surface becomes a rounded depression. Above it, the flame is convex-shaped and even on all sides. It's apparent that air currents are rising evenly and unimpeded from

the depression on the candle top, shaping the fire into the normal double-convex candle flame shape we expect to see.

2.3 Candle flame

2. Now turn the candle on its side as shown in the flame shape **diagram** 2.5 so air is not supplied equally to all sides of the candle. The shape is no longer uniformly convex and rounded. The flame assumes different shapes—convex, concave, or even straight, depending on the obstructions and impediments to air flowing to the candle. With practice, it is possible to become quite adept at producing flame shapes that resemble pennants, sails, or squares by cutting the candle to change the air flow as shown in **diagrams 2.6** and **2.7.**

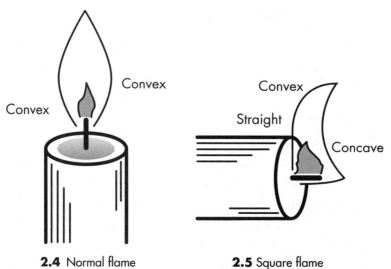

2.4 Normal flame   **2.5** Square flame

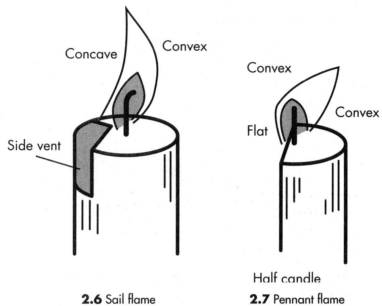

2.6 Sail flame　　　**2.7** Pennant flame

How to Ignite Smoke

www.Video1.ThePracticalPyromaniac.com

www.Project1.ThePracticalPyromaniac.com

Materials

- Long-handled matches
- Taper (candle) in candleholder

Light a candle and then gently blow it out, disturbing the air surrounding the wick as little as possible. You'll notice a wispy trail of white smoke rising from the wick. If you hold a lit match two or three inches from the wick, you will observe a trail of fire flashing downward through the air until it relights the candle. But you must do this quickly, or else the vapor condenses to liquid or becomes too diffuse to ignite.

2.8 Lighting candle smoke

What does the ability to relight the candle by igniting the smoke tell us? It means that the smoke contains combustible vapor. The vapor came from the candle wax that was turned to a gas by a physical process called vaporization occurring within the flame.

This leads us to another question: where in the interior of the flame does the vaporization occur?

- - - - - - - - - - -

Exploring the Interior of a Flame

www.Project1.ThePracticalPyromaniac.com

Materials

- ⊙ Propane torch with flame-spreading tip
- ⊙ Fireplace matches or long-handled lighter
- ⊙ Leather or other heat-resistant gloves
- ⊙ Safety glasses
- ⊙ (1) piece Pyrex glass tubing, 4 mm OD (outside diameter) × 2.4 ID (inside diameter) × 0.8 mm wall thickness, 12 inches long
- ⊙ Heatproof surface
- ⊙ Small clamp
- ⊙ Candle

Make an L-shaped Glass Tube

1. Affix a flame-spreading tip to the propane torch and then light the torch.
2. Wearing gloves and safety glasses, rotate the piece of Pyrex glass tubing in the hot part of the torch flame for several seconds until it becomes soft, as shown in **diagram 2.9**.

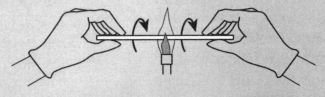

2.9 Heating glass tube

3. Remove the glass tubing from the heat and bend it into an L-shape with the desired angle, as shown in **diagram 2.10**.

2.10 Bending glass tube

4. Place the bent glass on a heatproof surface until it cools.

1. Place one end of the cooled bent tube in the candle flame as shown in **diagram 2.11**. Use a small clamp to hold the tube in place.

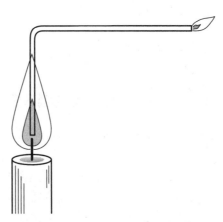

2.11 Placing end of cooled bent tube

2. Using a match or lighter, test the gas exiting the tube for flammability. Test various parts of the flame—the top, the bottom, the middle, and the edge—to see where the most flammable vapor is generated.

If you test the various regions of the flame carefully, you will find that the most significant production of flammable gas occurs in the center of the flame, just above the wick. The vapor produced in the middle of the flame can be carried through a tube and ignited at the other end.

You can see how the liquefied, burnable paraffin travels upward through the wick, where the heat in the candle interior vaporizes it. The paraffin is vaporized at the upper end of the wick and is present, unburned, in the interior space of the candle flame.

Making a Heat Map

Back in 1860, Michael Faraday performed this experiment for his audience, and it provides a great deal of insight regarding the interior of a flame. Given that this experiment is also simple, it's likely that most members of the audience duplicated the experiment as soon as they returned home.

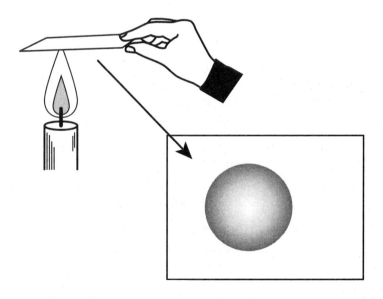

2.12 Heat Map

Materials

- Several sheets of medium weight (20 pound or greater) paper
- Candle
- Matches or lighter
- Bucket of water

1. Momentarily hold the sheet of paper inside (not over) a candle flame as shown in **diagram 2.12**. Part of the paper will turn brown. Keep careful watch and remove the paper before it starts on fire. If the paper ignites, extinguish it in the bucket of water.

2. It may take a couple of tries, but when done correctly, a ring-shaped burn mark is formed. This mark is a temperature map of the flame interior. The darker the color, the higher the temperature of the flame. The map shows that the highest temperature is at the edge of the flame, where the paraffin vapor meets the air. The shape and color of the darkened area on the paper is proof that the edge of the candle flame is "the heat factory" and the location of the most vigorous chemical reaction.

HOW FIRE CHANGES THE WORLD

A candle flame is a chemical reactor. We've figured out that the ingredients going into the flame are air and paraffin fuel. After the paraffin vaporizes in the center of the flame, it combines with components in air to produce heat at the edge of the flame.

The stuff going into the flame combines in the very special chemical reaction we call fire and produces by-products, namely water and carbon dioxide. Water is a product of the combustion of the candle. In fact, no matter what the fuel, from oil in oil lamps to the fluid in cigarette lighters, fire produces water.

Besides water, flames produce something else: carbon dioxide. Every time carbon fuels are burned—whether paraffin, coal, oil, or wood—carbon dioxide is produced. All day long and all over the world, millions of fires are burning. And every fire—in car engines, in home furnaces, and in power plants— takes fuel and oxygen and combines them to give off heat and the by-products of water and carbon dioxide. While the water produced is of little environmental concern, the carbon dioxide is a problem. It's the source of the greenhouse effect, which most scientists agree may have a serious detrimental effect on the Earth's climate. From these experiments we now know quite a bit more than we did at the start of the investigation about the nature of fire.

We know that to react, hydrocarbon fuel in the paraffin wax changes from solid to liquid to vapor, and that the wick is the key to moving the paraffin upward as it changes phase. The formation of the cup at the top of the candle and the way that airflow shapes the flame prove that oxygen in the air plays a very important role. Our tests for flammable vapor show that the fuel is vaporized in the center of the flame, just above the wick. The paper heat map clearly shows that the process produces high temperatures and that the highest temperatures are at the edge of the flame.

So as evidenced by the highest temperatures, the main chemical reaction occurs at the edge of the flame, at the interface between air and vaporized fuel. (In a later project we'll use an instrument called a thermocouple to measure this more precisely.) Finally, we learned that fire consumes stuff and makes new stuff from what it consumes. A flame is a chemical reactor that takes hydrocarbons and oxygen, produces water and carbon dioxide, and provides heat in the process.

Modern chemists use descriptive sentences made up of numbers and letters called chemical equations to concisely describe chemical processes. The equation describing the process going on within the flame of the candle is:

$$C_{25}H_{52} + 38O_2 \rightarrow 25CO_2 + 26H_2O + heat, where$$

$C_{25}H_{52}$ is the chemical formula for paraffin, which is a hydrocarbon fuel

O_2 is oxygen

CO_2 is carbon dioxide

H_2O is water

If you look carefully, you will find that these simple experiments verify each entity in the equation! You now know quite a bit about how the paraffin fuel combines with oxygen in the air to make heat. This process, or chemical reaction, is called oxidation. Oxidation continues as long as fuel and oxygen are supplied. We can say it's a self-sustaining reaction because we don't need to add any additional heat; the reaction makes more than enough of its own to continue.

So at last we arrive at a reasonable definition for fire, one that we'll use many times in this book:

> *Fire is a high-temperature, self-sustaining, chemical oxidation reaction of a hydrocarbon fuel resulting in carbon dioxide, water, and heat.*

Some may think that's too short, and others may find it a mouthful, but it's a definition that works. We'll explore this definition and its implications on the pages to follow. Michael Faraday would be proud.

- - - - - - - - - -

The Flame Tube

www.Video2.ThePracticalPyromaniac.com

In 1860, Pieter Rijke, a Dutch physics professor at Lieden University, was investigating the relationship between sound and fire. He took a large glass tube, stuffed a piece of iron mesh inside it, and held the tube over a gas flame until the mesh was red hot. Suddenly, the contraption emitted a long, sustained musical tone. The tone was so loud that office workers several rooms away complained to Rijke about the racket.

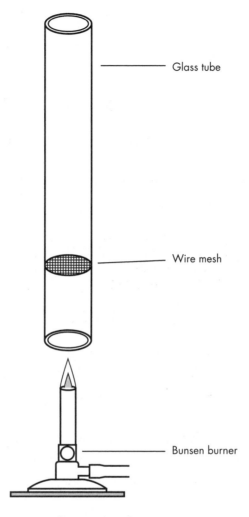

Glass tube

Wire mesh

Bunsen burner

2.13 Rijke's singing tube

Rijke's colleagues were intrigued with the device and set to work trying to discover the reason for "the singing flames." At first, scientists thought the sound was due to the periodic evaporation and condensation of water vapor produced by the fire. But Michael Faraday, by using a nonhydrocarbon fuel that produced no water by-products, disproved that theory. Building on this, later scientists showed that the sound was caused by waves of air, set in recurring motion due to the effect of the fire on the air's density. Hot air is less dense than cold air and moved upward, while the cool air sank. As the vibrating air moved through the tube, it resonated or vibrated at the natural frequency of the tube.

Several years later, a German scientist named Heinrich Rubens expanded on this idea. Rubens knew from Rijke's work that fire could produce powerful resonating waves of sound. Now he wondered if it was possible to use fire to make a visual representation of those sound waves. From the work of Faraday and other scientists, he knew quite a bit about the nature of fire and the motion of waves in fluids. The light and ethereal nature of flames, he figured, would be a nearly perfect medium for making sound waves visible for the first time. Setting to work in his laboratory at the University of Berlin, he came up with the wonderful device now known as the Flame Tube, which is often called the Rubens Tube in his honor.

KEEPING SAFETY IN MIND

1. This project involves open flame and uses propane. It is for adults or for those under the close supervision of an adult. The metal surfaces get very hot; do not touch them until they are cool.
2. Perform this project on a nonflammable surface and keep combustible objects well away from the flames.
3. Gas leaks are a possibility. Perform this project only in a well-ventilated space.
4. Keep a fire extinguisher handy.

www.Project2.ThePracticalPyromaniac.com

Materials

- ❯ (1) 2-inch-diameter iron or steel conduit, 5 feet long (You can use steel pipe as well, but don't use plastic pipe. Even if you protect the pipe surface with aluminum tape, the heat from the flame tube will soften and melt it.)

- Tape measure
- Marking pen
- Electric drill or drill press (While a handheld drill works, 200 is a lot of holes to drill!)
- $1/16$-inch drill bit
- $9/16$-inch drill bit
- $3/8$-inch 18 NPT tap (This is a pipe tap. It is *not* a $3/8$-inch hole tap.) You'll also need a handle to turn the tap.
- (2) push-to-connect plastic tube fittings: adapter $3/8$-inch OD (outside diameter) tube × $3/8$-inch pipe thread (Available in the plumbing aisle at hardware and home stores such as Home Depot.)
- Pipe thread compound
- (1) push-to-connect plastic tube fitting: adapter $3/8$-inch OD tube × $3/8$-flare fitting
- (1) push-to-connect plastic tube fitting: tee, for $3/8$-inch OD tube
- (1) roll $3/8$-inch OD vinyl tubing for fittings above
- Package of helium quality balloons (Helium quality balloons are thicker and less likely to leak than regular balloons.)
- Rubber bands
- (4) 2 × 4-inch wood pieces, 12 inches long
- (1) 2 × 4-inch wood piece, 30 inches long
- (1) box of 2½-inch deck screws and drill bit to fit
- Loudspeaker and amplifier (A small, monaural amp and a 3-inch speaker is plenty, although if you have larger old hi-fi equipment lying around, feel free to use it.)
- Optional: Pipe flashing boot (sometimes called a "witch hat") (This is a conically shaped piece of rubber designed to join pipes of differing diameters. The boot is an easy and secure way of attaching a larger diameter speaker to the 2-inch conduit or pipe. It is available at large home stores such as Home Depot in the roofing materials aisle.)
- Frequency generator and music sources (Free or inexpensive frequency generator applications are available on the Internet for personal computers, iPhones, iPads, and other handheld computing devices.)

- ❯ (1) propane regulator from high pressure or standard gas grill (There is a safety device in the regulator that shuts off the gas if you open the valve too quickly. Open the valve *very slowly*.)
- ❯ (1) 20-pound propane tank
- ❯ Safety glasses
- ❯ Long-handled lighter

1. Beginning 8 inches from the end of the conduit, make a series of marks with the marking pen ½-inch apart in a straight line extending across the top of the conduit. Stop marking 8 inches from the other end.

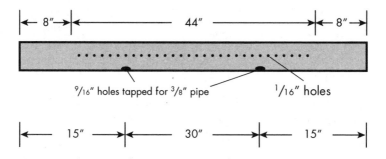

2.14 Flame Tube drilling diagram

2. Drill ¹⁄₁₆-inch holes on the marks, taking care to make the holes as perpendicular to the circumference of the conduit as possible. This will take a while, as there are nearly 100 holes to drill. To reduce drill wobbling, move the drill bit up into the drill chuck to shorten the exposed drill length.

3. Rotate the conduit 120 degrees and drill two ⁹⁄₁₆-inch holes 15 inches from each end. Again, take care to make the holes as perpendicular to the circumference of the conduit as possible. Tap the holes with the ³⁄₈-inch NPT tap. (To tap a hole in metal, place a drop of oil on the threads of the tap and then insert the tap into the hole. Turn the pipe tap clockwise three-quarters of a turn, then turn the pipe tap counterclockwise one-half of a turn. Continue this process until you are able to go all of the way through the hole with the pipe tap.)

4. Smear pipe thread compound on the pipe threads side of the adapter push-to-connect fittings and screw them into the two tapped holes until they bottom.

2.15 Flame Tube gas fitting

5. Next, plumb the gas supply. Insert tubes into the push-to-connect side of the plastic fittings on the conduit. Run 30 inches of plastic tube from each fitting to a tee push-to-connect fitting. Insert a 12-inch length of plastic tubing into the remaining open port of the tee fitting.

6. Insert the other end of the 12-inch tube into the push-to-connect side of the flare fitting.

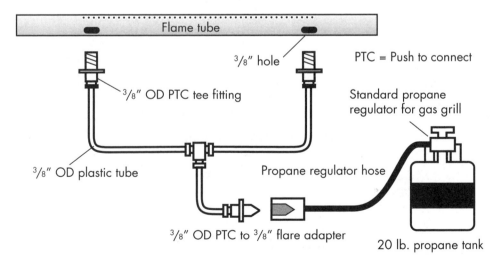

Flame tube

$3/8''$ hole

PTC = Push to connect

$3/8''$ OD PTC tee fitting

Standard propane regulator for gas grill

$3/8''$ OD plastic tube

Propane regulator hose

$3/8''$ OD PTC to $3/8''$ flare adapter

20 lb. propane tank

2.16 Flame Tube plumbing diagram

7. Cut off the narrow end of two balloons and attach one to each end of the conduit. Stretch them tight and secure each with a rubber band.

8. Build a simple stand from the wood pieces. Begin by placing two 12-inch long wood pieces into an X shape. Secure with two 2½-inch-long deck screws. Repeat to make a second X with the remaining pieces.

9. Connect the X brackets to one another with the remaining 2 × 4, using the deck screws.

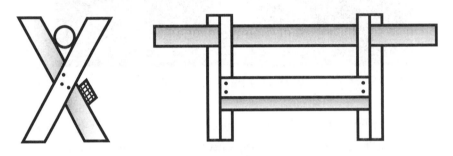

2.17 Flame Tube stand

10. Place the conduit on the stand with the holes pointing up. Make certain the area around the conduit is clear of combustible materials on all sides as well as above for at least 5 feet.

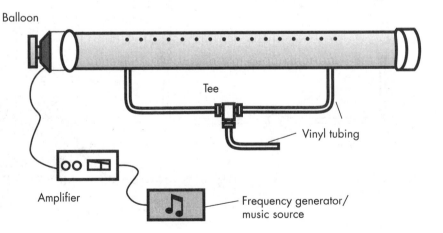

2.18 Flame Tube assembly diagram

11. Place the loudspeaker firmly against one of the balloons. The quality of the connection between the loudspeaker and the balloon has a large impact on how well the Flame Tube works. You may need to use duct tape or form a flexible gasket from rubber to affix the speaker to the end of the conduit to get a good connection. The use of the optional pipe flashing boot or witch hat is an easy way to get a good connection. Just use a knife to cut appropriately sized holes and couple the conduit to the loudspeaker.

12. Wire the speaker to the amplifier and frequency generator.

13. Attach the flare fitting to the propane regulator and propane tank.

14. Put on the safety glasses and slowly open the gas valve. Using the long-handled lighter, ignite the gas jetting out of the holes. Adjust the gas pressure so each gas jet is about ¾-inch high. The height of all gas jets should be equal. If they are not equal, turn off the gas and clean clogged holes by redrilling.

15. Turn on the frequency generator and amplifier and set it to 440 Hz. If you've done everything correctly, the frequency of the sound wave will be visible in the flame.

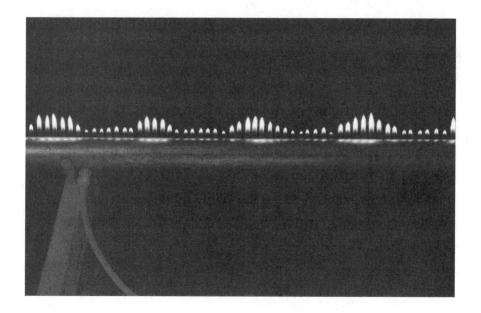

2.19 Flame Tube sound wave

The Science of the Flame Tube

1. Experiment with different frequencies to see different patterns and wave shapes. Replace the frequency generator with a music source and experiment with different types of music to understand how music affects the pulse and shape of the flames.

2. The Flame Tube is a wave form visualizer. It works because sound is a pressure wave, so as it moves through a gas like propane, it alternately compresses and expands the gas in different regions. When a constant tone of, say, 440 Hz (middle A) is pushed through gas via a speaker, a stationary wave—that is, a wave that seems to stay in a fixed position—is set up.

   The stationary wave causes areas of high pressure to appear at fixed points along the pipe, spaced half-wavelengths apart. Where the pressure is high, the propane is driven more strongly out of the pipe, resulting in a tall flame. Between these high-pressure points are low-pressure points, which create shorter flames.

⚠ **3** ⚠

THE FIRST LIGHTS

To ancient peoples, fire meant both heat and light. In the primitive societies of the Stone Age, light was prized, but the overarching need was the other component of fire, warmth, especially during the cold winter months when the heat of a fire meant the difference between life and death.

As society advanced, light came into its own. In almost all cultures, light and the devices that produced it were thought of as gifts of nature, and torches, lamps, and candles became symbols of divinity, life, and harmony.

In Africa and Europe, field scientists have uncovered the fossilized remains of campfire-charred bones so old that they likely predate *Homo sapiens*. Such archeological evidence suggests that our humanoid ancestors began taming fire perhaps as long as a million years ago. While protohumans most likely did not have the wherewithal to kindle fire, they did, it seems, have the mental capacity to capture naturally occurring fire, tend it, and preserve it for long periods.

About 20,000 years ago, in the Late Stone Age, humans painted rather elaborate images deep within several caves in Western Europe, the best known being the caves of Lascaux in southwestern France. Narrow and deep, these caves are impenetrable to daylight. The archeological evidence indicates that to produce the hundreds of artworks now considered some of the world's oldest, the painters must have manufactured the world's first artificial indoor lights as well. The experts postulate that the primitive Rembrandts most likely would have carried a burning firebrand from a campfire or have placed a few lumps of animal fat on a stone with a natural depression, then lit the fat with a burning faggot.

Soon after primitive humans domesticated fire, they invented the torch. It didn't take much of a technological leap to grab the nonburning end of a firebrand from the campfire and hold it aloft to spread light a small distance from the communal campfire. And doing so would certainly have empowered ancient humans to roam farther, reducing the real dangers of wild animals and treacherous terrain and the imagined ones of evil spirits lurking in the darkness.

Torch construction became more sophisticated over time. Early torch crafters bound sticks of resinous woods such as pine, yew, or juniper, wrapped them together with wet rushes, vines, and tendrils, and painted them with flammable substances such as resin or pitch. Later artisans constructed more sophisticated devices: hollow cases of clay or metal that they filled with pitch or tar and set aflame.

Although it is a simple device, the torch has a uniquely long-lived importance in human history. Perhaps nothing we still use today was invented so long ago. In some parts of the world, the fire from a flaming torch was the dominant source of illumination for 99 percent of human history.

Even today, from the Olympic Games to the Statue of Liberty, the torch remains one of the most important and significant symbols in cultures throughout the world. It remains an invariably positive symbol, representing life, hope, and goodwill.

Greek mythology tells the story of Prometheus, a Titan who stole fire from heaven and concealed it in a reed. As he ran back to earth as swiftly as he could, Prometheus swung the reed to and fro to keep alive the all-important flame. For this offense, the enraged Zeus chained Prometheus to a rock, where each day an eagle attacked him and ate his liver. Each night his liver regenerated, until Hercules shot the eagle and freed Prometheus.

The Greeks of the Golden Age expressed their gratitude by means of a festival in Prometheus's honor. What better way to honor him, they felt, than to recreate the actual deed by which the civilizing element of fire came to earth? The simple but appropriate tribute they instituted is called the festival of the torch-race, or *lampas*.

There is an excellent description of the *lampas* by Pausanias, one of the foremost travel writers of classical Greece. After his visit to Athens, Pausanias tells of a festival called the Promethia. He wrote, "There is an altar of Prometheus; from it towards the city, a race is run with burning torches. The point of the contest is to run swiftly yet keep the torch burning at the same time. If the torch goes out in the hands of the first racer, he loses the victory on that account, and then the next runner is declared the victor."

The festival of the *lampas* was a tremendous spectacle, especially when held on moonless nights. Anticipation would build throughout the day, and fires in Athens were extinguished before the race began. Finally, at the designated time, with all eyes looking skyward and nearly complete darkness enveloping the city, a lighted torch fell from a high tower, and the race began. The racers competed in the classical style, namely unclothed except for wreaths on their head.

Although they ran as fast as they could toward the center of the city, this was not a race of mere speed. The racers had to keep careful watch on the flames of their torches, tracking the swirling winds and their own slipstream and cradling the torches behind or in front of their bodies. In short, they had to do whatever was necessary to keep the torch from blowing out as they ran. When the winner crossed the finish line, he was feted like a god on Olympus by the admiring citizenry. Then the city's lights were rekindled from that runner's torch, signifying the city's renewal and purification.

The torches of the Greeks, Romans, and other Levantine peoples eventually gave way to oil lamps. But in some parts of the world, torch fire remained the principal source of illumination.

Most notably, northern Europe remained a torch-based society even through medieval times. The pine woods of Germany and the Baltics, thick with flammable, resinous trees, made torch light the most common way to light homes in Northern Europe. In fact, torch use predominated up to the time of the Renaissance. Historical research shows that even a rich Frenchman of the 14th century would go to bed by torchlight instead of oil lamps.

TORCHES OF POLITICS

Presidential political campaigns were much different in the 19th century than they are now, and to many political historians, they sound like much more fun. Instead of televised debates and commercials, scripted sound bites, and endless media analysis, the key political tool was the parade.

While everyone may still love a parade, Americans of 150 years ago, it seems, were enamored of them. Imagine for a moment you are a member of the "Wide Awakes," one of many political marching clubs organized to drum up support for political candidates. It's a pleasant summer night, and word has been received that a march on behalf of your presidential candidate, Abraham Lincoln, has been organized. This is terrific news! Since marching is what you like to do, you and your fellow Wide Awakes do it often and are very good at it. Everyone in the group (and there are thousands) owns a torch. Your torch—a new gimbal-mounted, nickel-plated tin torch in the shape of a Union Army musket—is particularly eye-catching.

In the evening, the Wide Awakes, as do all political marching clubs, wave their torches with pride and artistry, even using them in the manner of rifles presenting a display of close order drill to the crowds lining the streets. It's very exciting:

> Thousands of torches flashing in high, narrow streets, crowded with eager people and upon house-fronts in which

every window swarms with human faces, the rippling, running, sweeping and surging sounds of huzzas from tens of thousands, with the waving of banners and moving transparences of endless device are an imposing spectacle and these everyone in the city saw at the Wide Awake festival on Wednesday night.

—*Harper's Weekly*, October 13, 1860

Parades often lasted two to three hours. The costumed or uniformed participants sang campaign songs and shouted slogans as they marched.

To satisfy the need for the thousands of torches that accompanied such parades, scores of small manufacturing companies sprang up across the United States to fabricate parade torches. They made torches in many shapes and forms, ranging from rifle look-alikes for the aforementioned close order drill ceremonies to torches built in the shapes of faces, animals, letters (*L* for Lincoln), hats, pinecones, brooms, and pickaxes.

Torchlight parades as a political campaign tool peaked during the 1876 presidential election between Rutherford B. Hayes and Samuel Tilden. It was a boom time for torch manufacturers. Their factories ran at full steam, stamping out hundreds of thousands of unusually shaped torches for the closely contested election. Night after night, all over the country, people marched by torchlight, hoping the bright lights held aloft would awaken sympathetic feelings in onlookers and carry their candidate to victory.

But the era of such campaigning tactics was soon to wane. In 1876, strategies such as parades were the best way to reach people of all social statuses, literate or not. But as literacy rates rose and newspapers became less politically biased, political campaigning became less spectacular and more educational. By 1900, the importance and frequency of the torchlight parade declined dramatically, and the torch manufacturing industry slid into a steep decline from which it never recovered.

- - - - - - - - - - -

How to Make a Parade Torch

No angry mob in a 1930s horror movie would think of chasing down a monster or dispensing with a werewolf-infested old mill without a blazing torch in hand. Here's how to make your own torch.

www.Project3.ThePracticalPyromaniac.com

Materials

- ◉ Drill with $5/8$-inch bit
- ◉ (1) empty and clean metal 2.6-ounce Sterno or similarly sized can with a push-on lid
- ◉ JB Weld or other high-temperature epoxy adhesive
- ◉ (1) $5/8$-inch hex nut
- ◉ (1) piece of ½-inch-diameter cotton rope, 2½ inches long (Alternatively, you can braid together narrower diameter cotton ropes to make one thicker piece.)
- ◉ (1) 1-inch-diameter wooden dowel, 3 feet long
- ◉ Aluminum foil or high-temperature aluminum tape
- ◉ Kerosene (Do *not* use gasoline or alcohol.)
- ◉ Fill spout for the kerosene
- ◉ Long-handled lighter or fireplace match
- ◉ Fire extinguisher

1. Drill a $5/8$-inch hole in the center of the lid of the Sterno can (or similar clean can).
2. Using JB Weld or epoxy, glue the hex nut over the hole as shown in **diagram 3.1**. Be sure to let the JB Weld or epoxy harden before using your torch. Check label directions for curing time.

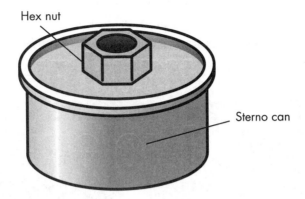

3.1 Torch fuel supply

3. Insert the piece of rope through the hex nut in the can lid. It should fit snugly with about a ½-inch wick sticking out as shown in **diagram 3.2.**

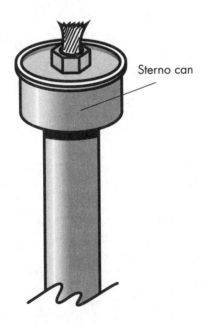

3.2 Torch assembly

4. Using JB Weld or high-temperature epoxy, glue the can to the wooden dowel.

5. Fill the can one quarter to one third full with kerosene, put the lid on the can, and press it down firmly. There should be about an inch wick of rope sticking out.

6. Cover the can with aluminum foil, forming a skirt around the can and the hex nut on top with just the rope wick sticking out. See **diagram 3.3**.

3.3 Completed Parade Torch

KEEPING SAFETY IN MIND

1. Use the torch outdoors only.
2. Kerosene is not as flammable as gasoline, but extreme caution is still required. It must be stored in an approved container.
3. Keep a fire extinguisher handy. Use extreme caution when lighting, handling, filling, or holding the torch. Never fill the torch while it is hot.

USING THE TORCH

1. Fill the torch with kerosene outdoors using a fill spout.
2. Make certain the lid is securely on the Sterno can after filling.
3. Let the rope wick draw kerosene up. After one to two minutes, light the wick using a long-handled lighter or fireplace match.
4. Do not hold the torch at an angle or it will drip kerosene.
5. If desired, you can whittle the other end of the dowel to a point so the torch can be placed in the ground in your backyard.

- - - - - - - - - - -

As human culture progressed, so did the use of fire to provide illumination. Lamps were invented in the Late Stone Age, the era from roughly 30,000 years ago to about 6,000 years ago. At a time long past, the details now unknowable, some unidentified genius observed that a twisted skein of fibers could transport liquefied fat or oil from the depression of a flint-knapped stone reservoir to the upper end of a fiber wick. He or she further discovered that if the tip of the wick were set afire, the result was a bright, long-lasting light.

The invention of the lamp was an important step in the progress of civilization, providing humans with the ability, not to mention the motive, to stay awake after the sun had set. This opened the door to a far richer domestic life. Families could now stay awake as late as seven, eight, or even nine o'clock! They probably spent much of that time inside homes, now illuminated by lamplight, by talking, playing, and no doubt fabricating new lamps and other domestic articles.

The stone lamp was a distinct improvement over the torch. It was easier to light, it was far easier to carry, it could be set down, and it did not litter the ground with ashes and hot embers as it burned. Stone lamps of Neolithic origin have been found from England to the Mediterranean and from to Russia to India. However, because they were time consuming and difficult to make, they were eventually displaced by ones made from easier-to-fabricate materials. Ancient peoples proved adept at fashioning lamps from materials at hand. In marine areas, lamps were made from seashells and chalk. Inland they were made from bone and sandstone and were fueled by whatever naturally burning, organic substance was locally available. In the far north, it was whale blubber. In parts of the Middle East, lamps were fueled by petroleum products such as liquid asphalt and naphtha collected from seeps in the ground.

Oil lamps were used continuously for millennia. In fact, because they are so useful and so simple, few major changes or improvements in the technology have been made. In most ways, the stone lamps of the Lascaux cave dwellers are the same as those of mid-18th-century England and America: they gave off the same amount of light for the same amount of oil. Twenty thousand years is a long time for an object to exist without a major improvement in technology!

Long ago, before the ubiquitous availability of flashlights and other portable electric lighting equipment, the ability to whip up an oil lamp from stuff at hand was mighty useful. Benjamin Franklin, perhaps the most resourceful and ingenious person America has ever produced, made observations and conclusions regarding the nature of fire that were notable in many ways. In 1762 Franklin was sailing to Madeira, a small but important Portuguese island in the middle of the Atlantic Ocean. While at sea, Franklin's dark, cramped cabin needed more light. With little trouble, the resourceful scientist produced a brightly shining lamp from materials aboard ship.

> During our passage to Madeira, the weather being warm, and the cabbin windows constantly open for the benefit of the air,

the candles at night flared and run very much, which was an inconvenience. At Madeira we got oil to burn, and with a common glass tumbler or beaker, slung in wire, and suspended to the ceiling of the cabbin, and a little wire hoop for the wick, furnish'd with corks to float on the oil, I made an Italian lamp, that gave us very good light all over the table. The glass at bottom contained water to about one third of its height; another third was taken up with oil; the rest was left empty that the sides of the glass might protect the flame from the wind.

Today the lamps most frequently depicted as coming from ancient times are those that were formed from fired clay and burned olive oil. African and Levantine lamps had open tops and were often hung on chains from the ceiling. Later, great numbers of Roman lamps were manufactured using molds instead of hand-forming techniques. They are among the earliest examples of mass-produced housewares.

Roman lamps had covers and sometimes multiple spouts and wicks, and such sophisticated devices provided considerable light. It was in the orange-red glow of burning oil lamps that people like Aristophanes wrote, Socrates philosophized, and Archimedes invented.

The lowly oil lamp, seemingly simple and utilitarian, is actually an incredible antigravity machine. Think about it. Why should oil, a substance obviously heavier than air, rise unaided from a reservoir up through the fibers of a wick to the tip, where a steady flame can be kindled until the bowl is emptied? What forces inside the fibers are stronger than gravity?

Although some of the improved 18th-century mechanical lamps, such as the Argand and the Carcel, are slightly different in terms of the shape of the wick, most oil lamps work the same way. Hydrocarbon compounds in the oil rise through the wick via a phenomenon called capillary action.

Capillary action arises from two different sources. The first is called surface tension or cohesion. It describes the attraction of molecules to molecules of similar kind. Cohesion is the atomic-level mechanism that causes oil molecules to follow one another, like army ants on a march, up the wick. The second process is adhesion, which is the molecular attraction between molecules of different types. Adhesion is the reason oil is attracted to the fibers of the wick in the first place. Together, these two processes make up the capillary action that seemingly defines the laws of gravity.

Designing and fabricating a simple olive oil lamp is easy and fun, and quite possibly useful. Best of all, when you make one, you form a connection with the technology of the past—not just the recent past, but the earliest times of human civilization. What your mobile phone is to you, the oil lamp may well have been to the cave dweller. The earliest lamps, called saucer lamps, were merely open bowls of oil with a wick placed along the edge. Eventually, the saucer lamp was superseded by the covered lamp, which had several advantages: it was less likely to spill, it usually had molded handles to make it easier and safer to transport, and its cover prevented contaminants from entering the oil reservoir.

The Olive Oil Lamp

An oil lamp is basically a reservoir with a support that holds a wick upright and a separate hole for adding the oil. Making a lamp on a potter's wheel is a simple task, as you need only throw a simple bowl, then pinch the wet clay to form a spout for the wick. You can also make a decent lamp by shaping clay with wet hands.

www.Project3.ThePracticalPyromaniac.com

Materials

- ⦿ 1 pound waterproof air-dry clay (Not all air-dry clays become waterproof when cured. For nonwaterproof clays, the lamp

42

interior may be coated with varnish or sealant if necessary to prevent oil seepage. Follow the directions on the package to cure and harden the clay. *Don't* use synthetic clays. They are not fireproof.)

- Waterproof varnish or glaze (necessary only if you are using nonwaterproof clay)
- Pencil
- Cotton buffing cloth (optional)
- Scribes or knives for decorating the lamp body (optional)
- Sandpaper (optional)
- (1) Piece of 100 percent cotton fabric ¾-inch wide × 4 inches long, or ³/₈-inch-diameter cotton rope
- 2 ounces of olive oil
- Scissors

MAKING A COVERED LAMP

1. Begin by forming the clay into a circle. At one side, pinch two pieces toward each other to form a spout as shown in **diagram 3.4**. If your clay is not waterproof, coat the interior with varnish to prevent the oil from seeping through the porous clay.

3.4 Forming lamp base

2. Next, shape an inverted bowl from damp clay and attach it to the open top of the saucer. Form a spout to match the one on the bottom, making sure it is open for the wick. Form a simple handle and attach it to the lamp.
3. Poke a fuel-filling hole in the top of the lamp with a pencil.
4. You may improve the finish of the lamp by lightly buffing it with cloth. The lamp may be detailed with scribes or knives or sanded if desired. The lamps of antiquity were often decorated. Motifs included mythological figures, animal and plant life, and abstract designs.
5. Make a wick. Use a piece of cotton rope or braid or twist a piece of cotton fabric into a tight wick.
6. When the clay has dried, fill the lamp with olive oil. Saturate the wick with olive oil and position it to extend from the bottom of the oil lamp to approximately ½ inch above the wick spout. Cut off the excess with scissors.
7. Light the wick and enjoy the warm, soft light. You may need to trim the wick at intervals to make it burn faster or slower depending on the amount of light you want it to produce.

3.5 Oil Lamp

KEEPING SAFETY IN MIND

1. This project is for adults or for those under the close supervision of an adult.
2. Olive oil is flammable. Avoid spills. Use the lamp with care to avoid fire danger. As always, ready access to a fire extinguisher is recommended.
3. Use this lamp outside on a nonflammable surface. Olive oil produces a beautiful soft, orange flame but generates a considerable amount of soot and smoke. Carefully choose the location in which you use the oil lamp to avoid getting soot on walls and ceilings. Oil lamps also may set off smoke detectors if used inside.
4. Handle the lamp with care.

4

THE ONE-CANDLEPOWER ENGINE

Bell, book, and candle; candle, book and bell; Forward and backward, to curse Faustus to hell.

—*Doctor Faustus*, Christopher Marlowe, 1604

After the lamp was well established, a new method of providing illumination was developed: the candle. While there is some evidence to suggest that ancient Egyptians may have used candles made from beeswax, most historians believe that candle making on a commercial scale was developed in Imperial Rome.

The Roman candle was a real innovation. It was made of tallow surrounding a cotton or flax wick. Tallow is made from fat that comes from cows or sheep. Since the average steer yields more than 100 pounds of fat, a great many candles can be made from each slaughtered animal. Candles were manufactured in more or less the same fashion for the next 1,500 years.

According to a candle making handbook of the 18th century, the best candles were made from "half Sheep's Tallow, and half

Cow's, but those made from Hoggs give an ill smell and a thick black smoak." To make evenly burning candles, says the handbook, the candle maker, or chandler, began by hacking big pieces of sheep and cow fat into easy-to-handle chunks. Next, the fat was placed in large iron vats under which a large fire was roaring. The fat, once fully rendered, would boil, and the chandler could skim the dross from the top of the vat and remove the larger pieces of crud by pouring the liquid through a screen. What was left was called tallow.

Next, wicks were lowered into the vats of molten tallow and then raised. The tallow adhered to the wicks. After the candle cooled a bit, it was dipped again, and a thicker coating of tallow was built up. The dipping and raising process continued until the candle was deemed to be the correct size.

Today the most common type of candle is made of paraffin wax. Harder and less smoky than tallow, paraffin is a by-product of the crude oil refining process. Commercialized by Scotsman James Young in 1850, paraffin is a white, odorless, tasteless, waxy solid with a melting point between 110°F and 150°F.

Candles (and to a lesser extent oil lamps) play an important role in the rituals of many religions. The symbolism of light plays a key role in many Jewish rituals, for example Sabbath candles and the menorah. Buddhist shrines are typically alight with many candles, placed carefully as a sign of respect for Buddha. Many Protestant religions integrate candles into their services to signify a host of beliefs, from the advent of Christmas to the resurrection of Jesus. Perhaps no religion incorporates candle symbolism to such a great extent as Roman Catholicism. Candles are used at every mass and church service as well as to show respect at the shrines of saints and to mourn the dead at funerals.

Among the most dramatic uses of candles is the rite of Bell, Book, and Candle, an ancient and darkly interesting excommunication ritual of the Roman Catholic Church. Although the phrase may conjure humorous images from a 1958 Jimmy Stewart comedy film, it describes a distinctly unfunny ritual of separation and cursing for apostates and heretics. In medieval times a bishop would perform this most solemn and disturbing rite. Holding a holy book and a bell, he would silently enter the church with 12 priests, each holding a lighted candle. The churchmen damned the heretic to the fires of hell with the utterance of these words:

> We declare him excommunicate and anathema, we judge him damned, with the devil and his angels and all the reprobate, to eternal fire until he shall recover himself from the toils of the devil and return to amendment and to penitence.

One of the priests would then ring the bell to symbolize a death knell, and another would slam the Book of Gospels shut. All 12 priests answered the bishop's statement with "Fiat [so be it]!" The priests would then blow out their candles, throw them to the ground, and dash them under their feet. The sinner would be left, according to the church, to an eternity of spiritual darkness.

INSIDE THE CANDLE FLAME

In an earlier chapter, we experimented with the mechanics of candle fire and demonstrated that (1) fire requires fuel and air, (2) fire produces carbon dioxide and water, and (3) the reaction begins at the wick and continues with more vigor as it moves toward the hottest edge of the flame. Now it's time to look closely, even microscopically, at the dynamics of fire, again using the candle as an example.

Paraffin wax, analytical chemists tell us, is composed mostly of moderately complex hydrocarbon molecules, a combination of 20 to 40 carbon atoms and a bit more than twice as many hydrogen atoms bonded together in long, snaky molecular chains. When heat from a match is applied, the paraffin molecules start to move with greater energy. When enough energy is applied, the paraffin molecules break loose from their neighbors in the wax, turning into a liquid thin enough to be sucked up into the wick via capillary action.

The high heat of a match head supplies enough additional energy to shake loose some of the paraffin molecules at the tip of the wick. Having now changed to vapor, the gaseous paraffin continues to heat in the match flame until so much energy is absorbed that the paraffin's molecular structure breaks down as the chemical bonds holding the individual atoms in each molecule are ripped asunder. In so doing, the vapor changes from a cloud of long, complex paraffin molecules into a mist of simpler but far more reactive fragments of carbon and hydrogen.

This process, called pyrolysis or "cracking," is the hidden midstep between candle wax and flame, the secret transformation that wood, coal, candle wax, or indeed, any burnable fuel goes through before it chemically breaks down to create the heat and light of fire.

The hot mist of reactive fragments spreads outward, soon colliding with oxygen in the air just beyond the immediate area of the wick. Now combustion takes place in earnest; carbon and hydrogen atoms quickly recombine to more and more stable forms until all that remains are (chemically speaking) the rock-solid molecules of water and carbon dioxide.

The oxidation process starts with long, energy-rich, but unstable molecules of paraffin wax and culminates in small, energy-poor water and CO_2 molecules. Along the way, the fire reaction produces heat and light.

Four general mechanisms occur simultaneously to produce fire at the top of a candlestick. Each mechanism resides in a particular area, or zone, of the candle. The first process takes place in the paraffin liquefaction zone, where heat levels are sufficient to turn solid paraffin into liquid that is transported up the wick. The area close to the tip of the wick is the fuel pyrolizing zone, where the fuel, in this case paraffin, turns from liquid to a gas or vapor, not yet burning. Once the fuel is vaporized, it combines with oxygen from the surrounding air but does so incompletely, because the oxygen concentration near the wick is fairly low. The incomplete combustion in the luminous flame interior, or inner combustion zone, forms soot and generates orange-colored light.

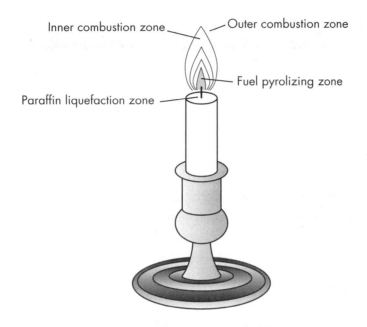

Inner combustion zone · Outer combustion zone

Fuel pyrolizing zone

Paraffin liquefaction zone

4.1 Four zones of a burning candle

The fuel vapors continue to move upward and outward from the luminous inner combustion zone and come into contact with oxygen in higher concentrations. This is the flame edge, or the outer combustion zone, where the reaction proceeds with the most vigor and highest temperature, but little light.

— — — — — — — — — —

While candles are now mostly important for their symbolic, ritual, and recreational uses, they were once more important for their practical ones. For hundreds of years, they served as the main source of light after sundown in homes and businesses. Candles have other uses as well. The One-Candlepower Engine project illustrates how a candle's heat can be used as well as its light.

Build a One-Candlepower Engine

www.Video3.ThePracticalPyromaniac.com

An engine, strictly defined, is any motor that converts thermal energy to mechanical motion. Although this project demonstrates perhaps the simplest thermodynamic engine imaginable, if you peer under the hood, there's quite a bit of interesting science.

KEEPING SAFETY IN MIND

1. This project is for adults or for those under the close supervision of an adult.
2. Perform this project on a nonflammable, easily cleaned surface or a surface that has been covered with aluminum foil for protection.

www.Project3.ThePracticalPyromaniac.com

Materials

- (1) 9- to 10-inch long candle of uniform diameter
- Knife
- (2) drinking glasses of the same size, at least 6 inches tall
- Ruler
- Pin or nail
- (2) matches or lighters

1. Expose the wick on the bottom of the candle by cutting wax away with the knife.
2. Trim the wicks on both ends of the candle so they are the same length.
3. Invert the drinking glasses and place them next to one another on the nonflammable work surface.
4. Using the ruler, find the midpoint of the candle. Heat the pin or nail using a match flame or lighter and insert it into the midpoint of the candle so it extends an equal amount on both sides.
5. Place the pin so it balances on the two inverted drinking glasses as shown in **diagram 4.2.**

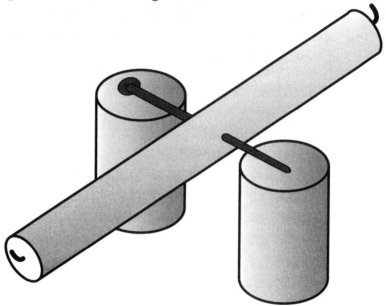

4.2 One-Candlepower Engine

6. Light both wicks simultaneously.

Shortly after the candle is lit, it will start to seesaw on the pin or nail pivot, slowly at first and then quite dramatically. The candle engine is fun to watch and, theoretically, the motion of the candle can be made to do some work, such as generating electricity or lifting water.

Why should our solid fuel engine behave this way? The rotating motion is due to a continuing serial weight imbalance, causing first one side and then the other to dip and rise. As we saw in chapter 2, the shape and geometry of the wax under the flame affect the shape and burning rates of the flame above it.

Once the candle begins to tip, the relative orientation of the flame to the wax candle changes, and along with it, the size of the surface area of wax in close contact with the flame. This affects the burning rates for both ends of the candle and therefore how much wax is consumed at any moment. This changing variable leads to an inequality in the weight of the candle first on one side, then on the other of the pivot. The system constantly self-adjusts, trying to regain equilibrium by burning more wax on the lighter side. This causes the seesaw motion until either the wax is consumed or the burning ends reach equilibrium.

⚠ 5 ⚠

THE FIRE DRILL

"Many-sided men have always attracted me," Franklin Delano Roosevelt told an interviewer in 1932. "I have always had the keenest interest in five men." Roosevelt's pantheon of genius consisted of Thomas Jefferson, Benjamin Franklin, Napoleon Bonaparte, Theodore Roosevelt, and Benjamin Thompson.

It is only a small overstatement to say that our understanding of fire and combustion revolves around the life and work of Benjamin Thompson, who became known as Count Rumford. Of all the characters associated with the advancement of the understanding of fire, likely none led as fascinating a life as Thompson. He had several careers in his colorful life: farmer, soldier, spy, social architect, aristocrat, and scientist. He succeeded fantastically in nearly all.

Benjamin Thompson was born to a poor family in Woburn, Massachusetts, in 1753. At 19 years of age, Thompson, although bright and clever, did not seem especially destined for greatness. He was slim and athletic, with a refined, gentlemanly appearance, and the happy possessor of a fine head of hair, a straight nose, and handsome, piercing eyes. He was also ambitious and personable. After moving from Woburn to Rumford (now Concord), New Hampshire, he met a wealthy young widow, Sarah Rolf, who was the daughter of the city's most influential citizen, the Reverend Timothy Walker. With charm and manners that well complemented his looks, Thompson soon won Sarah's heart. Just four months later, they married.

Thrust suddenly into the position of wealthy farmer and land-owner, young Thompson proved adept at social climbing, success-fully making his way into New England's landed classes. His efforts culminated in an introduction to New Hampshire's governor, John Wentworth, who was duly impressed by Thompson's bearing, his horsemanship, and above all, his profound self-confidence. Gover-nor Wentworth appointed him major in the Massachusetts colony's 2nd Provincial Regiment.

The governor's spur-of-the-moment appointment was a bit of a mixed blessing. Thompson was a complete unknown within the colonial military establishment, and he was young to boot. His com-mission stunned the regiment's older, more experienced officers. To put it mildly, they did not warmly embrace the upstart newcomer. To make matters more difficult, this was the time just prior to the American Revolution. Since Thompson had been close friends with many of the British Army officers and Tory administrators in Bos-ton, he found himself at the center of much controversy.

In 1774 a committee of Concord citizens charged him with "being unfriendly to the cause of liberty." Although he was acquit-ted of the charge, Thompson endured an atmosphere ranging from distrust to outright hostility. He was charged with and tried for treason and espionage a second time. Again, he was acquitted. But doubts as to his loyalty remained. Were these doubts justified, and was Thompson in reality a British spy?

In 1950, Sanborn Brown and Elbridge Stein, two highly trained document analysts and experts on invisible inks, examined a well-preserved letter Thompson had sent to British general Thomas Gage in 1775. The letter's content seems innocuous, even boring, at first glance, full of humdrum niceties and little else. But under chemical analysis, a 700-word secret message appears, providing the reader with detailed military intelligence about a planned American rebel attack on a redcoat stronghold in Boston. The letter also expresses Thompson's vow to "dedicate my life and fortune to my sovereign, King George the Third."

Thompson, being the spy that modern forensics proves he was, wisely decided that his second acquittal was likely to be his last. He quit Boston for England in March 1776. Once there, Thompson began his scientific career in earnest, taking a position in a military laboratory. He was particularly interested in gunpowder and ballistics. Using precise experimental techniques, Thompson determined optimal methods for preparing, storing, and using gunpowder. At the time, gunpowder was the most important chemical on Earth, because using it well was the key to battlefield success. Thompson's breakthrough work was of such value to the British government that on the basis of this contribution alone he was elected at the age of 26 as a fellow of the Royal Society, the most prestigious scientific institution of the era.

Thompson's charm, good looks, and sterling scientific reputation firmly established him within British society, and his star continued its rapid ascent. In 1783, the ruler of Bavaria, Carl Theodore, became aware of this British/American polymath's reputation. He offered Thompson the position of Major-General of Calvary and Privy Counselor, a position of great influence in both Bavaria's military and its administrative government.

5.1 Count Rumford

Thompson moved to Munich and set to work tackling Bavaria's governmental and social problems. At the top of his list was the miserable status of its army. Thompson must have been distraught after his first few troop inspections. The Bavarian military was, bluntly put, a mess, consisting of a poorly paid, ill-clothed, and badly equipped rabble, commanded by far too many corrupt and arrogant officers. Morale among the soldiers was low and performance worse.

Thompson instituted major reforms, removing unneeded officers and increasing soldiers' pay. By offering free education to soldiers and their families and engaging them in public works projects to occupy time not consumed in training, Thompson was able to transform what had been a mob of slobs into a far more effective military force.

Thompson instituted similar reforms among the poor on Munich's dirty streets. It is estimated that at the time of Thompson's appointment, nearly 1 in 20 Bavarians subsisted by begging in the streets. These beggars weren't simple street people with hands outstretched, hoping for a few coins; they were professional gangs with attendant turf wars, fights, and social ills of all types.

On New Year's Day, 1790, Thompson's newly improved army flooded Munich's streets, rounding up the beggars from their roadside perches and placing them in workhouses, which Thompson called "houses of industry." A believer in both carrots and sticks, he encouraged the beggars to work in exchange for decent food, housing, and medical care. His social programs reduced corruption and raised the standard of living throughout the country.

Carl Theodore marveled at Thompson's success and in appreciation for his work bestowed upon him the title of Count of the Holy Roman Empire. Thompson became Count Rumford, bearing the name of the New Hampshire town where his spectacular career had begun.

In his personal life, Rumford was a complex, multifaceted character. He could be overbearingly arrogant even while devoting his energy to the betterment of the lower classes. He was in equal parts and simultaneously duplicitous and loyal, scheming and trustworthy. But his contributions to society far outweighed his interpersonal shortcomings.

— — — — — — — — — —

After Bavaria, Rumford returned to London, where he conceived and promulgated the idea for "the formation in the Metropolis of London, a Public Institution for diffusing the Knowledge and the general Introduction of useful Mechanical Inventions and Improvement." Rumford was paid a call by professional philanthropist Sir Thomas Bernard, who admired Rumford's work improving many practical household devices. Bernard was even more taken by the man himself. Bernard and Rumford, feeding off one another's strengths and interests, came up with the idea for the Royal Institution.

Rumford and Bernard knew that many scientific ideas that could help people of all social statuses were never given the attention they deserved. Upon much reflection, they decided that there was a critical need for a forum where good scientific ideas could be discussed and, if worthy, commercialized. In late 1799 the two men proposed to their influential friends and acquaintances that they form a society called the Royal Institution. It was to be organized for the purpose of "diffusing the knowledge and facilitating the general introduction of useful mechanical inventions and improvements" throughout Great Britain. The idea caught fire, so to speak. Officers were elected, and managers were appointed to run the affairs of the organization.

To fund the Royal Institution, they organized a subscription plan through which each patron who contributed 50 guineas to the organization received the title of Perpetual Proprietor, and those who gave less were given a lesser title. There was great enthusiasm among London's philanthropic elite, and soon money started rolling in—more than enough to start the operations of the organization in 1799.

The Royal Institution on Albemarle Street grew to contain a research laboratory for resident scientists, meeting rooms, and a public outreach program, made famous by Michael Faraday's brilliant public lectures, including "The Chemical History of a Candle." The members of the Royal Institution, their friends, and their colleagues are inextricably linked to and responsible for our modern understanding of fire.

As much as Rumford enjoyed his time at the Royal Institution, he didn't stick around long to enjoy what he had achieved. Ever ambitious and restless, he soon left London for France, where he met, courted, and married Antoine Lavoisier's widow just a few years after the great French scientist met his demise on the guillotine. It was not a happy marriage, as their personalities were nearly polar opposites. In a letter to his daughter by his first wife, Rumford wrote, "Madame de Rumford and myself are totally unlike and never ought to have thought of marrying . . . I call her a female Dragon!" They fought in private and in public, culminating with their own domestic version of the War of the Roses, during which the ex-Madame Lavoisier poured boiling water over Rumford's prized flowers. Such a marriage could not and did not last. Rumford left his wife and house in Paris and took up housekeeping in suburban Auteuil, where he lived the rest of his life.

Between his many intrigues Count Rumford was responsible for a number of important technological advancements, including the drip coffeemaker, kitchen oven, and pressure cooker, as well as thermal underwear, central heating, the smokeless chimney, and numerous other useful items.

The greatest contribution Rumford made to science was not one of his many inventions but something more basic: the role he played in determining the true nature of heat. It is because of Count Rumford's seminal discoveries and his contribution to the basic understanding of physics that scientists such as Humphry Davy, Michael Faraday, and Rudolf Diesel could make their own significant contributions to science and the understanding of fire.

THE NATURE OF HEAT

No book on fire could be considered complete without exploring the fundamental nature of heat. At the time Rumford lived, not much was understood about the nature of heat. While everyone innately understood that some objects feel warm and others feel cold, distinguishing between what was heat and what was fire was a difficult proposition.

In the late 18th century, there were two competing theories regarding the nature of heat. The first was the caloric theory of heat and held that an invisible thing or fluid called "caloric" was contained within warm objects. The more caloric something contained, went the theory, the more heat it contained and the hotter it would feel. The second theory said that heat was not a thing but rather a vibration of some sort. According to this way of thinking, the faster the particles making up a body vibrated, the more heat it contained. This was called the kinetic theory of heat, and it was the theory to which Rumford subscribed.

Starting in 1798, Rumford carried out a series of experiments in an effort to determine which theory was correct, caloric or kinetic. Did heat result because an invisible substance, namely caloric, was released from its hiding place inside an object? Or, did the external application of movement to an object cause the object to vibrate and thus heat up?

In the armory in Munich, Rumford immersed a brass cannon barrel blank in water and began boring a hole in the muzzle using a large boring tool turned by two horses driven in a circle. The machining operation generated so much heat that nearly 20 pounds of cold water could be boiled within two and a half hours of boring. What's more, Rumford showed that the supply of frictional heat was inexhaustible as long as the horses kept the boring tool turning.

Heat came not from the release of a hidden fluid inside the brass, stated Rumford, but from friction caused by the motion of the boring bar against the brass cannon barrel. If heat was produced by caloric, at some point the caloric would be used up and no more heat would be produced. Since the heat supply was unlimited, as long as friction continued, Rumford concluded that heat had nothing to do with invisible substances. Heat, he wrote in his notebook in large capital letters, "is MOTION."

- - - - - - - - - - -

Fire in the Hole: Making Fire from Friction

While Count Rumford formally proved that heat comes from motion, primitive peoples had a deep, hands-on appreciation of making fire from motion—the motion of rubbing sticks together. The fire drill is a simple but ingenious device for making fire from friction, and it does so in a (just slightly) less laborious fashion than the proverbial rubbing of two sticks.

There are numerous ways to make fire that are far easier, faster, and less punishing on the hands than using fire bows, fire ploughs, or fire drills. So why should anyone take the time to use such primitive methods to light a campfire? As Isaac Watts taught, learning is best done experientially, for it is by firsthand observation that the brain best absorbs lessons. While you can read all you want about aboriginal fire-starting techniques, unless you put down the book and pick up the sticks, you'll never really gain an appreciation for the process. It is far more delicate and precise than most people imagine. Great care is required in selecting the wood, gathering the correct type of tinder, and constructing the fire drill itself.

Making a fire by friction using the motion described by Rumford is difficult, but the feeling of accomplishment gained by persevering is great. It's an activity I recommend every true student of technology and science undertake at least once.

www.Project4.ThePracticalPyromaniac.com

Materials

- ❯ Flywheel (see Step 1)
- ❯ Electric or hand drill

- Drill bits
- Pencil
- (1) 1 × 2-inch pine board, 22 inches long (pushing handle)
- (1) ¾-inch-diameter oak dowel, 3 feet long (spindle)
- Medium sandpaper or belt sander
- Screws and nails
- Epoxy
- Screwdriver
- Pocket knife
- (1) 1 × 4-inch cedar board, 12 inches long (hearth board)
- (1) ¼-inch nylon or polyester cord, about 56 inches long
- (1) 1 × 2-inch pine board, 3 inches long (palm board) (optional) [see Tips and Troubleshooting]
- Tinder*

*A note on tinder: Tinder consists of very dry and fine fibers. Selecting tinder materials is very important. Not every thin, woody material can be used. Cedar bark makes very good tinder bundles, as does the dried inner bark of poplar and cottonwood trees. Peel off the dry bark and work it with your fingers to shred it into fine pieces. Dry grass will work, but it can be difficult to ignite. Dryer lint, especially if collected from clothes made of 100 percent cotton, is frequently used. One very dependable type of tinder is called oakum. Oakum is made from jute fibers, which is the material gunnysacks are made of. Traditional wooden ship builders use oakum as a type of waterproof caulk by pressing it into the seams of planks that make up boat hulls.

You can purchase a length of oakum from a shipbuilding supply store and fluff it up into a small bundle of easy-to-ignite tinder. Devotees of primitive fire-starting techniques often pull apart the fibers of a gunnysack. No matter what type of tinder is used, it must be dry and light, with plenty of surface area to allow easy ignition and continued burning.

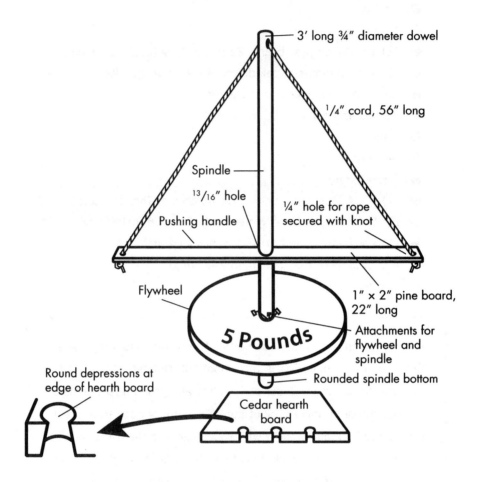

3' long ¾" diameter dowel

¼" cord, 56" long

Spindle

¹³/₁₆" hole

Pushing handle

¼" hole for rope secured with knot

Flywheel

5 Pounds

1" x 2" pine board, 22" long

Attachments for flywheel and spindle

Round depressions at edge of hearth board

Rounded spindle bottom

Cedar hearth board

5.2 Fire Drill

1. First you need to construct your flywheel, which can be made in a number of ways and from many materials. Aboriginal peoples and primitive fire-starting enthusiasts have used everything from rocks to coconuts to make a flywheel. The only requirement is that it possesses a large "moment of inertia," which is the combination of size and weight that makes a body, once spinning, want to keep spinning. A 5-pound barbell plate works well. Alternatively, a 12-inch-diameter,

1-inch-thick disk cut from plywood and loaded with additional weight can be used.

2. Drill a ¾-inch hole in the exact center of inertia of the flywheel. You can locate the center fairly accurately by finding the point on which the flywheel comes close to balancing when you set it on a pencil. Get as close as you can to the balance point, and mark it.

3. Drill a $^{13}/_{16}$-inch hole in the middle and a ¼-inch hole in each end of the pushing handle, as shown in **diagram 5.2.**

4. Drill a ¼-inch hole 1 inch from the top of the dowel or spindle.

5. Round off the top and bottom of the spindle with sandpaper or a belt sander. While the top should be well rounded, leave the center of the spindle bottom fairly flat, or blunted, to maximize friction with the hearth board.

6. Mark a point 4 inches from one end of the spindle. Attach the flywheel to the spindle at that point. You will need to improvise (use screws, epoxy, nails, etc.) to get a solid connection.

7. Use the knife to make a round depression about half the diameter of the spindle at a point approximately $^{1}/_{8}$ inch deep and about half an inch from the edge of the hearth board. This is "the hole," from which the frictional heat will be created and fire will be generated. As shown in the diagram, the hole is open to the side and bottom to allow air circulation.

8. Put the dowel through the hole in the center of the pushing handle. Thread one end of the cord through the hole at one end of the pushing handle and tie a knot on the underside of the handle, the side nearest the flywheel. Thread the other end through the hole at the top of the dowel, and then down through the hole at the other end of the handle. Knot it in the same manner as the other side.

9. Place the spindle on the depression ("the hole") on the hearth board.

10. Spin the handle around the spindle as shown in **diagram 5.3**.

USING THE FIRE DRILL

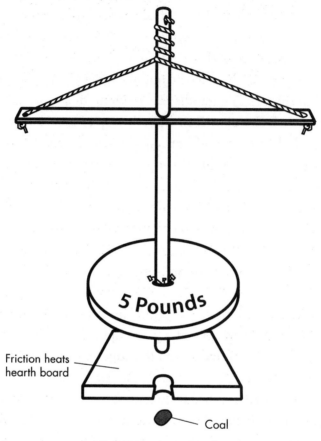

5.3 Using the Fire Drill

Now the fun begins. Apply a smooth but firm push down on the handle. After you apply the downward motion, the flywheel cause the spindle to continue turning, and the cord wraps around the spindle. When the push handle moves back up the spindle, the fire drill is ready for the next push, which causes the spindle to rotate in the opposite direction. It may take you a little while to get the rhythm of it.

As the pushing handle bobs up and down, the spindle will grind against the hearth board, producing a larger hole. Once the hole has begun to form, stop pushing and use the knife to whittle a V-shaped notch in the side of the hearth board, extending about

1/8 inch into the hole. See the detail inset on the bottom left of **diagram 5.2**.

Now continue to pump the handle up and down for as long as it takes for the fire drill to heat up the hearth board to ignition temperature. Eventually a hot ember, called a "coal," will form in the edge of the hole on the hearth board.

With the knife blade, quickly and carefully remove the hot coal. Place it on the tinder nest, then blow gently until the tinder ignites. There, you've made fire from motion!

TIPS AND TROUBLESHOOTING

Producing fire from friction is difficult and falls more under the category of art than science. It takes a great deal of experience to become proficient at starting fire via friction, and it is common for beginners to be unsuccessful in their first several attempts. Even though the directions here describe a method with which I've often had success, your results may be different. But persistence and good work pay off, and eventually will be rewarded with a glowing coal that can be blown into a roaring fire.

If you have trouble, consider the following points:

1. The wood must be dry.
2. The type of wood used in the hearth board is very important. Cedar is often used, but cottonwood, poplar, and yucca work as well.
3. The amount of heat generated is a function of the weight of the drill as well as the amount of time spent pumping the handle. Therefore, you may need to add weight to the flywheel to generate enough friction to produce a coal.
4. You can also add weight by using a "palm board," a small, palm-sized piece of hardwood. Carve a depression in the center of the palm board with the knife and grease it lightly. Ask a helper to place the depression on the top of the spindle and press down firmly on the palm board as you pump up and down on the pushing handle.

6

THE BURNING RING OF FIRE

Why did so many people leave Europe for the New World starting in 1620? A good case can be made that the image that really
sold America in the minds of wistful potential colonists was that of
roaring fires in their fireplaces.

Beginning in the 1600s, most of the colonists and other settlers
who made the arduous journey across the Atlantic did so because they
were drawn by the promise of America's abundant natural resources.
Certainly the prospect of owning land—something nearly impossible
for those not lucky enough to be firstborn sons—made the wild, harsh
forests and fields of Massachusetts and Virginia very attractive.

Throughout England and the rest of Europe in the 17th and 18th
centuries, wood for cooking and heating was in chronically short supply for the lower and middle classes. Like today's oil prices, the price
of firewood went up and down, but mostly up. Over time it doubled,
doubled again, and then tripled in many communities. As the price
of wood rose, more poor and middle-class people spent their winters
uncomfortable at best and freezing at worst. Some desperate souls
resorted to stealing wood from richer people's hedges and fences.

So, when the British heard early reports from the New World
of endless forests of "oak, ash, elm, willow, birch, beech, pine,

and fir" for the taking, it seemed like a wonderful place indeed. Francis Higginson, an early Massachusetts colonist, wrote in 1680 that while the winters of New England were "more sharp than is in old England," that was not a problem for him as "We have plenty of fire to warm us, a great deal cheaper than in London." Gloating perhaps a bit, he claimed that a poor servant in America may have a greater fire than a nobleman in Europe. "Surely, here is good living for those that love good fires," he wrote.

The prospect of easily and inexpensively staying toasty warm in winter was among the foremost reasons that multitudes of settlers made the voyage west. In their mind's eye they saw a better life and, more specifically, a fire blazing in a cozy hearth in a rough but pleasant house surrounded by small but fertile fields.

The colonists were lucky that wood was so plentiful, because the technology for burning it had not advanced far since caveman times. Heating a house was difficult because fireplaces were terribly wasteful. The best fireplaces, which belonged to the richest families in America, were immense but thermodynamically crude affairs. They could be the size of a modern walk-in closet, up to three and half feet deep, five feet high, and ten feet wide. The walls were straight, and there were no dampers or doors, making them inefficient heating devices and gluttons for fuel. These cavernous compartments ate up firewood by the cartload.

It's not surprising, then, that by 1750, the once abundant supply of firewood, in New England at least, was gone. While not approaching the stratospheric prices in England and the rest of Europe, firewood was no longer cheap. Benjamin Franklin wrote that firewood now "makes a very considerable article in the expense of families" and that "wood which for the last 100 years might be had at every man's door must now be fetched 100 miles to some towns."

Franklin, a brilliant inventor as well as politician, writer, and philosopher, set to work on an improved fire-burning device. The fruit of his labor was a cross between a closed stove and an open fireplace that came to be called the Franklin stove.

First made in 1742, the Franklin stove was much different and much more efficient than any previous design. It had an open front like a fireplace, but it stood in the center of the room like a stove. And unlike a fireplace, the Franklin stove's cast iron walls radiated heat for a long while after the fire went out.

Its remarkable efficiency came from its ability to utilize all three methods of heat transfer—conduction, convection, and radiation—in a more advanced manner than any previous heating appliance. The improvements stem from special features that Franklin designed, such as sophisticated flues to minimize smoke and a honeycomb of internal chambers to warm air before directing it into the room.

Those who bought the stove, declared the brochure used to sell it, would find their rooms warmed evenly and efficiently so "people need not crowd around the fire" any longer and there would be no more problems with people "being scorched in front and froze behind." Such good, even space heating would reduce many of the diseases commonly occurring in winter including "colds, coughs, catarrhs, toothaches, fevers, pleurisies, and many other diseases."

Although Franklin could have controlled the rights to his invention and made a great deal of money on it, he declined to do so, saying, "As we enjoy great advantages from the inventions of others, we should be glad of an opportunity to serve others by any invention of ours; and this we should do freely and generously." Many stove makers began to manufacture the Franklin stove; they made improvements and refined it into the device still commonly used today.

There was a competitor to the Franklin stove on the other side of the Atlantic. Benjamin Thompson, otherwise known as Count Rumford, was born just 10 miles up the road from Franklin's home in Boston. Although there is no record of America's two greatest colonial scientists ever meeting, there is little doubt that they were familiar with one another's work and reputation.

In 1795, Count Rumford was hard at work on the problem of designing fuel-efficient stoves and fireplaces. When he cast his ever-observant eyes upon the fireplaces he found in the finest homes, he saw devices that had undergone even less improvement than had stoves. Eighteenth-century fireplaces were not much different than those built in the 12th century. They were basically masonry boxes set against an outside wall with a hood placed atop the box to collect and channel smoke through a wall and into a chimney. Rumford observed that the turbulent flow in the chimney resulted in smoky, dangerous downdrafts. He also calculated that seven-eighths of the energy of the fuel consumed by traditional fireplaces was carried up into the atmosphere in the smoke.

Such waste was intolerable to the sensibilities of the frugal and industrious nobleman. As a first remedy, he proposed that the sharp angles where the chimney met the hearth box be rounded or streamlined to improve the flow of smoke and gas upward. Next, he drastically narrowed the entrance from the hearth box to the chimney by adding what he termed a "throat," or a second vertical wall. Lastly and most importantly, he invented a movable door, now widely known as a chimney damper. Positioned inside the flue, the damper regulated the pace of a fire by controlling the airflow.

The genius in all this is that the fireplace and chimney, instead of simply allowing the smoke to waft upward in a lazy, uneconomical drift, now forced the smoke to virtually shoot up through the damper and out the chimney. Rumford's design was a tremendous improvement, resulting in far less smoke and extracting twice as much heat from the wood as the medieval designs that preceded it.

Unlike Franklin, Rumford had no qualms about making as much money as he could from the improved design. He publicized his fireplaces by publishing cartoons and poems that extolled the virtues of his new product. And he was successful. Thomas Jefferson had a Rumford fireplace installed at his home at Monticello. Henry David Thoreau wrote in his book *Walden* that Rumford fireplaces, along with plaster walls, copper plumbing, and large dry cellars, were among the most important comforts taken for granted by civilized human beings. Soon, thousands of Rumford fireplaces were in use and were competing with Franklin stoves for primacy in the hearth and home marketplace.

But Rumford wasn't done designing equipment for controlling heat and fire. He went on to invent a number of cooking devices, including the drip coffeemaker, the double boiler, the commercial kitchen range, and the gigantic and capacious "Rumford Roaster," all of which revolutionized the way fire is applied to food.

- - - - - - - - - - -

The Burning Ring of Fire

www.Video4.ThePracticalPyromaniac.com

> I fell in to a burning ring of fire
> I went down, down, down and the flames went higher.
> And it burns, burns, burns
> The ring of fire.
>
> —Johnny Cash, "Ring of Fire"

This practical project incorporates the work of Franklin, Rumford, and a later scientist named Thomas Graham. Franklin's and Rumford's stove-building concepts come together in an easy and useful project: a portable camping stove called the Burning Ring of Fire.

Many hikers on long treks, such as the Appalachian Trail or Pacific Coast Trail, make their own camping stoves. They do so to save money, to help the environment, and, most significantly, to prove to themselves that they can make something so useful and so scientifically elegant.

A good quality, high-performance camping stove can cost anywhere from $40 to $200, but you can make your own for practically nothing using a couple of metal soup cans, some odd bits of aluminum, and a nut and bolt. The Burning Ring of Fire uses widely available and inexpensive methanol fuel and has the advantage of being tough, light, and portable—but don't use it with cigarette lighter fluid, gasoline, or kerosene. It won't work, and the stove could explode.

Considering its cost and size, this stove is a good performer: the pressurized blue flame that this stove creates can boil a quart of water quickly and efficiently.

www.Project5.ThePracticalPyromaniac.com

Materials

- ➤ Hammer
- ➤ Sewing needle (size 5, although other sizes will work)
- ➤ (1) 18.8-ounce tin can (such as Campbell's Chunky Soup), lid removed, emptied, and washed
- ➤ (1) 18.5-ounce tin can (such as Progresso Soup), lid removed, emptied, and washed
- ➤ Hacksaw
- ➤ File or medium grit sandpaper
- ➤ Drill with $5/16$-inch drill bit
- ➤ (1) ¼-inch hex nut
- ➤ JB Weld or other high-temperature epoxy
- ➤ (1) plastic spoon
- ➤ Furnace cement
- ➤ Methanol fuel (such as Heet brand gas line dryer. Do *not* use gasoline or camping fuel.)
- ➤ Small funnel
- ➤ (1) ¼-inch-diameter bolt, $3/8$-inch long
- ➤ (1) 9-inch round aluminum pie pan
- ➤ Long-handled lighter or fireplace matches
- ➤ (2) bricks
- ➤ Pot, large enough to cover the stove
- ➤ Bucket of water or fire extinguisher

Refer to **diagram 6.1** while completing the following steps.

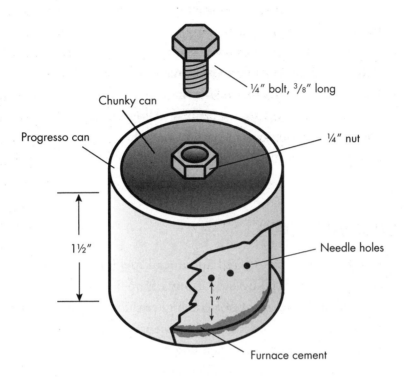

6.1 Burning Ring of Fire assembly

1. Use the hammer and sewing needle to poke 8 to 12 holes in the Chunky can, ½ inch from the closed end, as shown in **diagram 6.1**. The holes must be very small for the stove to work properly; nails and most drill bits are too large. The needle may bend or break when it is hit with the hammer, so use as little force as necessary to poke the holes. Shorter needles bend less than longer needles, so you may find it helpful to cut the needle in half.

2. Cut the cans to 1½ inches high using the hacksaw. Deburr the cut edge using the file or sandpaper. Carefully feel the edge to make sure it is no longer sharp and is safe to handle.

3. Drill a $5/16$-inch fuel filling hole in the center of the bottom of the cut Chunky can. Center the ¼-inch nut on the hole and

cement it into place using the JB Weld or other high-temperature epoxy. Take care not to get any cement on the threads of the hex nut.

4. Use the plastic spoon to coat the bottom of the Progresso can with a thick layer of furnace cement.

5. Center the Chunky can inside the Progresso can and press the open end into the furnace cement. Allow the cement to dry completely. Note: It is important that the cement seals the cans together with no leaks.

USING THE STOVE

1. Carefully pour the desired amount of fuel into the stove through the fuel-filling hole using a funnel. Do not overfill. The alcohol level must not rise higher than the gas jet holes on the Chunky can. One fluid ounce is a good start. Adjust upward or downward depending on the size of the meal being cooked. Once the stove is filled, insert the bolt in the fuel hole and tighten it to close.

2. Set the aluminum pie pan well away from vegetation, fabrics, and other flammable materials. Pour a few drops of the methanol fuel onto the aluminum pie pan on a flat, nonflammable surface. Close the methanol container and move it to a safe location.

3. Prime the stove by placing it in the middle of the small methanol "puddle" on the pan. With the long-handled lighter or fireplace match, ignite the methanol in the primer pan. Flames will envelop the stove (**diagram 6.3**) and heat the methanol inside, building pressure. After several seconds, the alcohol vapor will audibly jet out through the small holes, mix with air, and provide a hot blue flame in the circular ring between the two cans as shown in **diagram 6.2**. The stove should light from the priming flame, but sometimes it is necessary to light the gas between the cans with a long-handled match.

6.2 Burning stove

When the priming flame in the aluminum pie pan burns away, the Burning Ring of Fire's internal cooking flame will continue to burn. The stove is ready to use for cooking. Use two bricks or other sturdy, nonflammable supports for pots and pans as shown in **diagram 6.4**.

6.3 Priming flame

6.4 Using the stove

EXTINGUISHING THE STOVE

1. To extinguish the stove, deprive the stove of oxygen by covering it with a pot until the flame goes out.
2. This stove burns very hot. Allow it to cool completely before handling.

KEEPING SAFETY IN MIND

1. Camping stoves can be dangerous. The priming process produces a large alcohol flame that envelops the stove for several moments, so the stove must be used on a noncombustible surface. Do not use indoors, near flammable material, or inside a tent.
2. The alcohol flame can be hard to see in bright light. Always assume the stove is on until you can verify the absence of heat.
3. Make sure there is enough alcohol in the stove to cook your meal before igniting. It is dangerous to add fuel to a stove with a flame present.

- - - - - - - - - -

THE FUNDAMENTALS OF FUEL

One of Michael Faraday's colleagues, Scottish chemist Thomas Graham, published an important paper in the Royal Institution's scientific journal, the *Quarterly Journal of Science,* in 1829. It was entitled "A Short Account of Experimental Researches on the Diffusion of Gases through Each Other" and was one of the first scientific investigations into the way gases mix and combine. Graham continued his research, examining the manner in which gas molecules behave, and later developed the law for which he is best known: Graham's Law of Effusion.

Effusion is the process by which gas molecules escape from a container through a tiny hole. Graham found through careful experimentation that the effusion rate of a gas, that is, the speed at which a volume of gas jets through a small orifice, is proportional to the size of the gas molecule. Larger molecules effuse more slowly than small molecules. The exact proportion is equal to the square root ratios of the molecular weight. If that seems vexingly complicated, you need only envision a container full of mixed gases: the gas with the smaller, lighter molecules will stream out faster than the one with bigger molecules.

The reason the Burning Ring of Fire works with alcohol but not other fuels is due to the weight of the fuel molecules. The weight of air molecules is about 29 atomic units. The weight of methanol is about 32, and the weight of ethanol is about 46. So when the stove is primed, the air and alcohol vapor come out at a roughly 50/50 ratio and mix with more air, maintaining the Burning Ring of Fire at a size and temperature just right for boiling water. The molecular weight of gasoline is 110 atomic units, which is too heavy to jet out of the stove's small holes in the proper volume to sustain the smooth combustion desired.

7

The Hydrogen Generator and the Oxygenizer

There was nothing by way of wit, heroism, or humor in Henry Cavendish. Neither was there anything mean or dishonorable about him. Cavendish was simply one heck of a quiet fellow. And he was a scientist of the first order, whose discoveries played a central role in the modern conception of fire.

By present standards, you might say that Cavendish was something of a psychological basket case. He dressed in a manner that caused other scientists, not generally the height of fashion themselves, to shake their heads and chuckle. His taciturnity was legendary. His greatest peculiarity, however, was his extreme shyness.

To Cavendish, entering a room with people in it was a challenge. Colleagues told of instances when the man was stricken with shyness so severe that he would stand on the landing outside of the door, minutes ticking by, attempting to muster the courage required to actually turn the doorknob. When women were involved, the situation went from merely awkward to complete social meltdown. Cavendish was so irrationally fearful of women that he frequently could not bring himself to speak to them. His biographers marveled at the lengths to which he would go to avoid contacts with the fairer sex. He built a back staircase to his house simply to avoid encountering his female housekeeper, and he often communicated with his female servants by writing notes.

By all accounts, Cavendish's scientific skills in the fields of chemistry and physics were tremendous, but his abiding shyness made it nearly impossible for him to communicate, even with other scientists. As one colleague wrote, those who "sought his views speak as if into vacancy. If their remarks were . . . worthy, they might receive a mumbled reply."

Cavendish's contribution to our modern understanding of fire is significant. When he proved that water is made of two distinct gases and is not a chemical element itself, he opened the door that ultimately led to our understanding of oxygen and its central role in combustion and burning. Cavendish's work was the foundation on which Joseph Priestley and Antoine Lavoisier made key scientific advances. Cavendish is best remembered as the discoverer of hydrogen, which he isolated when he perfected the technique of collecting gases above water. He published a paper about his breakthrough techniques and the resulting findings in "On Fractious Airs" in 1766.

What are fractious airs? The man who observed and named them didn't really know either. A century before, English scientist Robert Boyle filled a flask with sulfuric acid and dropped small pieces of iron into it. Boyle inverted the flask in a larger container of acid then watched as bubbles of gas collected at the top of the inverted container. Boyle wasn't sure what this gas was, so he called it "fractious air" because he assumed it was some component, or fraction, of whatever comprised air.

When Henry Cavendish performed an experiment similar to Boyle's—also producing gas by dropping pieces of metal into acid—he had more advanced equipment and better laboratory techniques. Cavendish was able to subject the puzzling gas to a far more rigorous investigation. Among other findings, he noticed that the gas burned energetically, so he termed the gas "inflammable air." In actuality, the gas he was studying was hydrogen.

Quite taken with the stuff, he found he could produce it in a number of different ways. He found that his inflammable air was generated when hunks of iron, zinc, or tin were dropped into containers of muriatic (hydrochloric) acid or sulfuric acid. Interestingly, he also found that dissolving other types of metals in these and other acids did not produce the flammable gas.

Cavendish's great contribution to chemistry was that he proved that neither air nor water were elements, as had been believed for a thousand years. Each is a mixture of other, more basic elements. This finding raised the key question of what those basic elements might be. The ancient worldview that the world was composed of four irreducible elements—water, fire, earth, and air—was, due to Cavendish and his fellows, crumbling rapidly. In the next chapters, we see how this breakthrough led to the search for other elements, including oxygen. In fact, oxygen theory is the basis for the modern understanding of fire.

Cavendish performed more experiments, determining that inflammable air is very lightweight stuff, even compared to other gases. When French aeronauts heard about his discovery, they became excited, reasonably deducing that Cavendish's gas would be better filler for their balloons, as gas bags filled with inflammable air would presumably go higher and travel farther than balloons filled with hot air.

By 1783, less than 20 years after Cavendish's paper on fractious airs was published, French scientist Jacques Charles devised generation techniques so advanced that he could makes loads of hydrogen, enough to fill a manned balloon. Charles and his friend Nicolas Robert made the first untethered ascension of the skies in a basket underneath a gas hydrogen balloon on December 1, 1783, less than two weeks after the first manned hot-air balloon flight by the Montgolfier brothers in a Parisian suburb.

Another important and related contribution concerns Cavendish's discovery that water could be decomposed into two distinct gases. Cavendish's experiments, which were published in 1784, contain accounts of his investigations into the composition of water. This finding was of immense importance as it determined for the first time the fundamental makeup of water: that it is composed of two more elemental substances, and those substances are gases. The experiment proving this fact involved the observation that when hydrogen is combined with air and an electrical spark is applied, the mixture explodes and creates water.

— — — — — — — — — —

Henry Cavendish was among the first and most influential subscribers to the Royal Institution. Cavendish joined, at least in part, after becoming familiar with the work that Royal Institution founder Count Rumford had done on the nature of heat. Rumford's cannon-boring experiments were brilliant proof of what Cavendish already believed about the nature of heat and fire. The Honorable Mr. Cavendish became one of the Royal Institution's most active members, albeit in ways that required him to speak very little. His stature and money were vital in providing prestige and attracting funding for the fledgling organization.

Cavendish's work intertwined with that of many eminent scientists studying fire during the golden period from the late 18th century to the mid-19th century. Cavendish was on friendly terms with Count Rumford (to the extent that Cavendish was friendly with anybody), often meeting with him and other members of the Royal Institution for conversation and perhaps a glass of sherry as well. Cavendish and Benjamin Franklin were on quite good terms, and Franklin was much impressed with Cavendish's abilities as an experimental scientist. We know that Cavendish read Antoine Lavoisier's scientific papers and that Cavendish sent his papers directly to Lavoisier to read in Paris.

As a major benefactor of the Royal Institution, Cavendish was instrumental in bringing England's most famous scientist, Humphry Davy, to the Royal Institution as resident lecturer and researcher. When Cavendish died in 1810, Humphry Davy delivered a lecture on his work. "His processes were all of a finished nature, perfected by the hand of a master," Davy said. "They required no correction. Although many of them were preformed in the very infancy of chemical science yet their accuracy and beauty have remained unimpaired amidst the progress of discovery."

THE PNEUMATIC TROUGH

Around the time of Cavendish's earliest forays into chemistry in the mid-18th century, chemists began to wonder why a candle in a closed container extinguished itself rather quickly. Moreover, they wondered why a mouse placed in a closed vessel would eventually expire and why placing green plants in the same vessel could miraculously make the same air sustain mouse life once again. Many scientists spent a lot of time considering such questions. These good minds, with Cavendish at the forefront, eventually came up with the idea for a laboratory instrument that could help answer them.

Although it might not sound exciting, the history of the pneumatic trough is rather interesting. The pneumatic trough is one of those boring sounding, yet incredibly important scientific discoveries that provided an avenue for understanding a great deal of things that aren't boring at all. The trough, it seems, played a key role in building the foundations of the knowledge we now possess about fire.

Henry Cavendish's great advances in chemistry began with his experiments with the pneumatic trough. With it, he was able to isolate hydrogen, which he termed "inflammable air." This was the first of many gases to be discovered with the aid of the device. A pneumatic trough is a very important piece of laboratory gear, because with it one can isolate and collect a gas without contaminating it and store it securely in a jar or other container.

There are three main parts to the device:

The trough, typically a large glass tray filled with liquid
A gas bottle (or bulb) to hold the gas collected
A shelf to support the mouth of an inverted gas bottle with
the mouth of the bottle under the surface of the liquid.

The bottle is filled with water, inverted, and placed into a pneumatic trough already containing water. The outlet tube from the gas-generating apparatus is inserted into the opening of the bottle so that gas can bubble up through it, displacing the water inside the bottle with the gas to be collected.

The trough is important for undertaking the projects in this chapter and the next. The parts are easy to procure, and the project is easy to build.

- - - - - - - - - -

Constructing a Pneumatic Trough

www.Project6.ThePracticalPyromaniac.com

Materials

- ◉ Drill
- ◉ Aluminum sheet, 2 x 6½ inches
- ◉ Glass loaf pan, 1½-quart size
- ◉ 2 rubber bands
- ◉ 4-fluid-ounce jar (a standard spice jar works well) with straight sides and open top
- ◉ Water
- ◉ 2 feet of rubber tubing

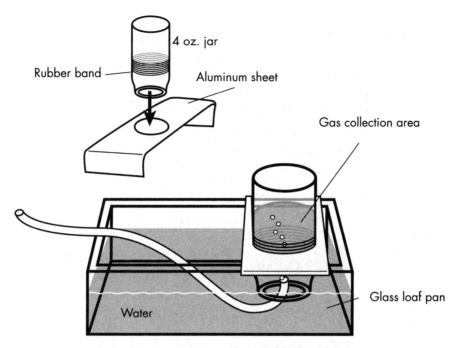

7.1 Pneumatic Trough assembly diagram

1. Drill a hole the same diameter as the jar in the center of the aluminum sheet.
2. Bend the aluminum sheet into a U-shaped bracket so it fits snugly over the width of the loaf pan.
3. Wrap the rubber bands around the middle of the spice jar and insert the jar through the hole in the aluminum sheet. The close-fitting hole in the sheet should allow the bottle to pass through, but not the rubber bands.
4. Add 1½ inches of water to the pan. Insert the rubber tubing into the mouth of the jar. Adjust the position of the rubber band on the jar so that the mouth of the jar is about an inch below the water surface.

KEEPING SAFETY IN MIND

The following projects involve the use of small amounts of acid and other chemicals and produce small amounts of flammable gas. Be careful to avoid spills. If you come in direct contact with the chemicals, immediately flush your skin with plenty of water for at least 15 minutes.

1. This is a good time to review the safety advice in chapter 1.
2. Be sure to read and follow all label directions.
3. Use eye protection at all times.
4. Do not breathe in chemicals; perform projects in a well-ventilated space.

- - - - - - - - - -

Hydrogen Generator

The Hydrogen Generator is a laboratory device that produces flammable hydrogen gas, and it is similar to the device Cavendish used. With the aid of a pneumatic trough, we are ready to tread the same path Cavendish took on the way to publishing his work on "inflammable airs" in 1766.

www.Project6.ThePracticalPyromaniac.com

Materials

- ◉ Safety glasses or goggles
- ◉ File or grinder
- ◉ (3) post-1982 "copper" pennies
- ◉ (1) 3-inch-long piece of 5-mm glass tubing bent into an L shape
- ◉ #6½ rubber stopper for flask, with a single 5-mm hole
- ◉ 8-inch square baking pan
- ◉ 150 ml water, plus cool water for the tray
- ◉ 250 ml flask
- ◉ Graduated cylinder for measuring the hydrochloric acid

- Rubber gloves (for use when handling acid)
- 50 ml of 31.45 percent hydrochloric (muriatic) acid (Chemists would call this a 10-molar solution. This is the concentration typically sold in hardware stores in plastic gallon jugs.)
- 2 feet of 4.8-mm-diameter rubber tubing
- Pneumatic trough (See the previous section for instructions on building the trough.)
- Leather gloves
- Taper candle, 6 to 12 inches long
- Long-handled lighter

Refer to **diagram 7.2** when completing the following steps.

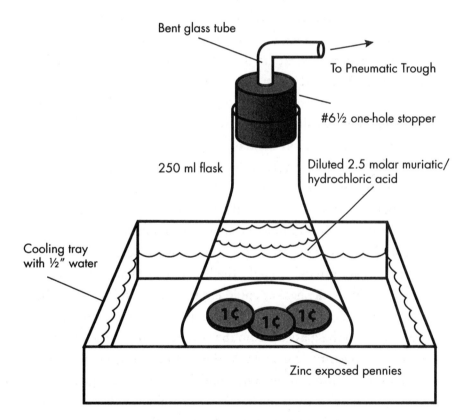

Bent glass tube

To Pneumatic Trough

#6½ one-hole stopper

250 ml flask

Diluted 2.5 molar muriatic/ hydrochloric acid

Cooling tray with ½" water

Zinc exposed pennies

7.2 Hydrogen Generator

1. Put on safety glasses. With the file or a grinder, remove as much of the copper coating from the pennies as possible, leaving a shiny white zinc surface. (U.S. pennies minted in 1983 and later are mostly zinc underneath thin copper plating. Check the date on the pennies before grinding.)

2. Insert the bent glass tube into the stopper. (For instructions on bending glass tubing, see the sidebar in chapter 2.)

3. Place the baking pan on a sturdy, nonflammable surface. Add about a half inch of cool water to cover the bottom. The water and tray prevent the reaction from moving too rapidly and the flask from becoming too hot.

4. Add 150 ml of water to the flask. Place the flask in the water-filled tray.

5. Put on rubber gloves. Carefully and slowly add 50 ml of hydrochloric acid to the water in the flask. Important note: Always add acid to water, never water to acid. Mixing water and acid generates heat, so if you add water to concentrated acid, it can boil and splatter dangerously.

6. Drop two or three zinc-exposed pennies into the flask. Hydrogen gas bubbles will immediately begin to form. Place the stopper from step 2 on the flask. See **diagram 7.2**.

7. At first, the gas issuing from the glass tube is a mixture of the air already inside the flask and the hydrogen from the chemical reaction. Because of the oxygen in air, this mixture burns explosively and must be vented into the atmosphere, so begin by allowing the initial generated gas to escape into the room. Do *not* attempt to light the hydrogen gas at the exit of the glass tube—the beaker could explode. After 60 seconds, the escaping gas is mostly hydrogen. Connect one end of the rubber tubing to the glass tube on the stopper and run the other end to the pneumatic trough. As the hydrogen is generated, it will displace the air in the bottle and fill it with hydrogen.

8. After a minute or two, the bottle will be filled with hydrogen gas. Change to leather gloves and remove the jar from the trough, keeping the open end facing downward.

9. Slowly lower the open end of the 4-ounce jar from the trough over a candle flame as shown in **diagram 7.3**. The hydrogen

will explode with a pop, and a small flame will be briefly visible in the mouth of the jar.

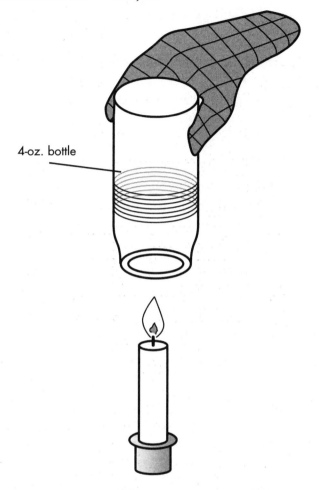

4-oz. bottle

7.3 Igniting hydrogen

A small hydrogen explosion like this is fascinating in and of itself. However, the hydrogen generator will be used to much more exciting effect in the following projects.

DISPOSING OF THE CHEMICALS

Dispose of unused hydrochloric acid by carefully pouring it down the toilet or laundry room drain. The reacted solution of acid and zinc pennies contains zinc chloride. Allow the solution to evaporate, and then wrap the remaining solid and dispose of it in the trash.

– – – – – – – – – – –

Joseph Priestley's life and personality were the opposite of those of the great men of fire we've already met. Whereas the scientific intelligentsia of England thought Henry Cavendish talked too little, they likely thought Joseph Priestley spoke far too much. And while Count Rumford began life as an American and ended it in exile in Europe, Priestley sailed in the other direction, starting as an Englishman and ending as an exile in America.

Still, Priestley's life had a number of important similarities with the lives of several scientific compatriots. Like Count Rumford, his opinions on social welfare and "betterment of the lower classes" were as important to him as his scientific work. Like Benjamin Franklin, Priestley was a polymath and a major contributor to many fields of human endeavor. Finally (what are the odds?), like fellow fire researcher Antoine Lavoisier, he wound up harassed and threatened by a mob of drunken political zealots. While Lavoisier literally lost his head, Priestly fortunately kept his.

Joseph Priestley, like Michael Faraday, Henry Cavendish, and many other scientists featured in this book, was schooled in Isaac Watts's teachings on experiential learning. He was also an amateur chemist of considerable ability. His great contribution to the understanding of fire resulted from a number of clever experiments he conducted using apparatuses of his own design, culminating in his 1774 announcement that "air is not an elementary substance, but a composition" of gases. Key among them was the colorless and highly reactive gas he called "dephlogisticated air," which the French chemist Antoine Lavoisier would soon name "oxygen."

The role oxygen plays in the reaction we know as fire may seem obvious now. But in 1774, discovering the existence of an odorless, colorless gas—much less understanding the role it plays in sustaining the process of combustion—was major scientific progress.

In one of his experiments, Priestley used a magnifying glass to focus the sun's rays on a bit of the compound mercury oxide placed inside a closed glass vessel. When he heated the compound to a high temperature, Priestly observed with amazement that it produced a gas in which a candle would burn more brightly and a mouse could live four times longer than in a normal atmosphere. Priestley, writing in the elaborate prose characteristic of his time, described the results of the experiments in a scientific paper in 1775:

> This air is of exalted nature. A candle burned in this air with an amazing strength of flame; and a bit of red hot wood crackled and burned with a prodigious rapidity, exhibiting an appearance something like that of iron glowing with a white heat, and throwing sparks in all directions. But to complete the proof of the superior quality of this air, I introduced a mouse into it; and in a quantity in which, had it been common air, it would have died in about a quarter of an hour; it lived at two different times, a whole hour, and was taken out quite vigorous.

Although Priestly could not accurately interpret these results using the scientific knowledge of the time, Antoine Lavoisier used Priestley's findings a few years later to construct the theories that now underpin modern chemistry in general and the understanding of fire in particular.

However, there is more to Priestley's story. In 1791, Joseph Priestley was a Unitarian minister in Birmingham, a large industrial city in the north of England. In the decades prior to 1790, Birmingham was a good place for those with less than orthodox adherence to the tenets of Britain's Anglican Church. Birmingham hosted a large and well-tolerated community of religious dissenters, including Quakers and Unitarians. But change was in the air, and those whose beliefs did not match those of King George III were starting to feel uncomfortable.

There were a number of reasons behind the change, ranging from small, local issues such as which books should be allowed in the local library, to larger controversies stemming from national arguments regarding support for the bloody revolution then under way in France. But perhaps the most volatile issue pertained to the rights of dissenting religious practitioners, that is, non–Church of England religious groups.

When it came to espousing his dissenting religious views, Priestley was not bashful. He spoke loudly and often and published several pamphlets in the years preceding 1791. Some, such as his provocatively titled "History of the Corruptions of Christianity," were considered dangerous and heretical by local leaders of the Anglican Church. A lightning rod for controversy, Priestley was nicknamed "Gunpowder Joe" by opponents who thought his views on church and state were, like gunpowder, both dangerous and powerful.

Anger at the dissenters in general and Gunpowder Joe Priestley in particular came to a climax on July 14, 1791. A drunken mob marched through Birmingham, setting fire to dissenters' property and their churches. At eight o'clock in the evening "a large and riotous number had again collected, and notwithstanding the attendance of the magistrates, demolished the windows in front of the tavern." No doubt helping themselves to a few bottles of alcohol, the rioters marched to their next targets. They burned down the Old and New Unitarian Meeting Houses and headed to the home of Joseph Priestley.

Priestley and his wife barely had time to flee the mob. For the next several days they hid in the homes of friends while the riots continued. Writing shortly after the event, Priestley described what he saw on the first night:

> It being remarkably calm, and clear moon-light, we could see to a considerable distance, and being upon a rising ground, we distinctly heard all that passed at the house, every shout of the mob, and almost every stroke of the instruments they had provided for breaking the doors and the furniture. For they could not get any fire, though one of

them was heard to offer two guineas for a lighted candle; my son, whom we left behind us, having taken the precaution to put out all the fires in the house, and others of my friends got all the neighbors to do the same. I afterwards heard that much pains was taken, but without effect, to get fire from my large electrical machine, which stood in the library.

The mob rioted for four days until finally being dispersed by soldiers who marched in from far-off Nottingham.

Soon after, Priestley resigned from his ministry in Birmingham, fearing that his presence would bring further harm to members of his congregation. Retreating to London, Priestley spent the next three years as a minister and teacher while publishing, among other works, the syllabus of the chemistry course he taught there. But Priestley could not find peace in London, either. His reputation, both scientific and theological, followed him. France and Britain declared war upon one another in 1793, and the British government began imprisoning suspected internal enemies. Priestley and his wife decided it was time to leave for America, to which their three sons had already emigrated.

While Priestley was likely most passionate about his metaphysical beliefs, he is best remembered for his transformational work as an amateur chemist. As we'll see in the next chapter, Priestley's research into the "doctrine of airs" (or in modern terms, the chemistry of gases) led directly to our modern understanding of the role of oxygen in combustion and fire.

Before the era of Cavendish and Priestley, air was thought of as just air. No one knew it was made up of multiple components as there was no concept of hydrogen, oxygen, nitrogen, and so forth. Even if someone had given thought to the idea that air was a mixture of other things, the tools of the day would have prevented any serious progress in dividing atmospheric air into its more basic components. Air, it seemed at the time, was nothing but air.

- - - - - - - - - -

The Oxygenizer

The oxygenizer is an oxygen-rich environment for conducting experiments that can be constructed from easy-to-find materials and chemicals.

> **www.Project6.ThePracticalPyromaniac.com**

Materials

- ❯ Newspapers, enough to cover your work surface
- ❯ (1) *non*alkaline C or D battery
- ❯ Hacksaw or rotary tool with cutting wheel attachment
- ❯ (1) small plastic spoon
- ❯ (1) small plastic or glass container with lid
- ❯ (1) pint-size Mason jar with metal lid (the screw ring is not needed)
- ❯ (1) 10-penny nail (approximate, other sizes will work too)
- ❯ Hammer
- ❯ Steel wool
- ❯ Graduated cylinder
- ❯ 25 ml of 3-percent hydrogen peroxide solution
- ❯ 75 ml water
- ❯ Cigarette lighter fluid
- ❯ Safety glasses
- ❯ Leather gloves
- ❯ Scale
- ❯ Long-handled lighter
- ❯ Fireplace matches
- ❯ ½ gram sulfur (optional)

Extracting Manganese Dioxide from a Nonalkaline Battery

1. Protect your working surface with newspapers, and then carefully cut off one end of a standard nonalkaline battery using the hacksaw or rotary tool with cutting wheel. Use caution, as the opening may be sharp. The moist, black powder in the battery (which is quite messy and stains clothing) is mostly manganese dioxide.

2. Using the small plastic spoon, remove the manganese dioxide from the nonalkaline battery and place it in the plastic or glass container with a lid. Remove the carbon rod for use in creating the Arc Light (chapter 10).

1. Poke a small hole in the center of the Mason jar lid using the nail and hammer.
2. Pull off a small piece of steel wool (about 1 gram) and unroll it so it is loose and "fluffy."

3. Roll one end of the steel wool into a tight tail and insert through the hole in the lid, as shown in **diagram 7.4**. The steel wool should hang down three-quarters of the way to the bottom of the jar.
4. Pour 25 ml of the hydrogen peroxide solution into the jar.
5. Pour 75 ml of water into the jar. Swirl gently to mix.
6. Squirt a drop or two of cigarette lighter fluid on the bottom end of the fluffed steel wool.

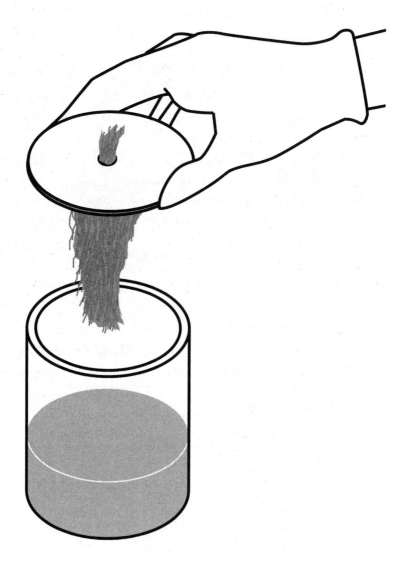

7.4 Oxygenizer

The next two steps should be done in rapid succession for best effect. Make sure your safety glasses are on.

7. Put on gloves. Place a 1-gram piece of manganese dioxide in the jar. The catalyzing action of the manganese dioxide is liberating oxygen from the hydrogen peroxide.
8. With a long-handled lighter, ignite the bottom the fluffed steel wool. The lighter fluid will catch on fire. Immediately and with great care, place the glowing steel wool into the Mason jar.

Once inside the oxygen-rich environment of the Mason jar, the steel wool puts on a glorious display of burning. The bright and vigorous mini-inferno is a wonderful, if short-lived, exothermic reaction and a fantastic demonstration of the reaction of oxygen with fuel that is central to the idea of combustion.

For an even more exciting demonstration, rub a small (about ½ gram) lump of sulfur in the fibers of the fluffed steel wool and stand back before igniting it. The intensity of the flame inside the jar is astonishing.

- - - - - - - - - -

Oxygen Re-ignition

This project uses the same materials as the preceding project, plus a fireplace match.

1. Fill the Mason jar with oxygen by again dissolving manganese dioxide in the hydrogen peroxide solution, using the same procedure as in steps 4, 5, and 7 in the preceding project. Light the long-handled fireplace match and blow it out. Before it stops glowing, place it in the Mason jar.
2. The match will roar back to life and burn extra brightly in the oxygen-rich environment. Oxygen combines with wood's hydrocarbons and produces the reaction we term "fire."

This is a good time to note the distinction between fire and burning. Steel wool, sulfur, and many other substances burn in an oxygen-rich atmosphere. The molecules combine rapidly with oxygen, giving off heat and light. They are correctly said to be burning. But are they on fire?

No. The precise definition of fire we developed early in this book is a high-temperature, self-sustaining, chemical oxidation reaction of a hydrocarbon fuel resulting in carbon dioxide, water, and heat. Thus, steel wool is not on fire because no hydrocarbon fuel is involved and no carbon dioxide is produced. However, by our definition, the fireplace match, being hydrocarbon-based, is on fire in the oxygen-rich Mason jar.

We now see how important oxygen is to starting and sustaining fire. In the next chapter we'll examine the life and works of Antoine Lavoisier, who is often considered the father of modern chemistry. He was the man who first explained the intricate chemical relationship between oxygen, burning, and fire.

△ 8 △

Exploding Bubbles

Antoine Lavoisier was an extraordinary man who led an extraordinary life and died an extraordinary death. He, in large part, laid the foundations of modern chemistry, allowing the scientists who followed him to determine the true nature of fire.

In 1794, Lavoisier was living comfortably in Paris, a man of enough leisure to study chemistry as a hobby in a well-outfitted laboratory that he set up in his home. He had trained as an attorney, but his money came from his position in the French government as Fermier-General, a special type of tax collector in the employ of the king of France. That was a very good job to have, at least before the French Revolution. In exchange for a fee paid to the royal government, Lavoisier, like his father before him, was given a monopoly on the sale of tobacco in a particular region of France. He bought low and sold high, and his fortune amounted to the modern equivalent of nearly $3 million.

With such wealth, Lavoisier could afford to quit his job and devote himself to the chemistry experiments that had become his passion. Lavoisier became famous throughout Europe for a great number of contributions to chemistry, including describing in detail the chemical reactions of what we now call combustion.

Lavoisier also developed the nomenclature modern chemists use to describe chemicals. The importance of this contribution cannot be overstated. Prior to the textbooks he wrote, there was no consistent, clear way to describe chemical substances or compounds. For example, if a man named Glauber or Epsom discovered a salt of some chemical interest, it was called "Glauber's Salt" or "Epsom's Salt." Substances were called by names as such sal-ammoniac, vitriol, and spirit of wine (now called ammonium chloride, sulfuric acid, and ethanol, respectively), none of which made sense in a larger context. The nomenclature system Lavoisier devised made scientific work and study much simpler and far more straightforward.

At the age of 50, Antoine Lavoisier was at the pinnacle of his chemistry career and seemed poised to go on to even greater achievements. Alas, the period of the French scientist's creativity and experimentation was about to end as the infamous Reign of Terror of the French Revolution unleashed political anger. Lavoisier wrote to his close friend and fellow scientist Benjamin Franklin in America, saying that he wished Franklin were in France to serve as an adviser and exemplar and perhaps to cool down the increasingly toxic political climate.

I greatly regret your absence from France at this time. You would have been our guide and you would have worked out for us the limits beyond which we ought not to go.

Indeed, limits were not a key part of the French Revolution. After each mass execution of the Revolution's "enemies of the Republic" (and there were many of them), throngs of fanatics, their fervor raised by the sight of the guillotine's work, paraded the streets. Often drunk and purposely dressed in the shabbiest of clothing, the mob, called the *"sans culottes,"* (meaning "those without fashionable pants") yelled *"A bas les rois!"* and *"A bas les aristocrats!"* (Down with the kings! Down with the aristocrats!)

When Lavoisier heard these street cries, no doubt he became quite troubled. His fortune from his position as an employee of a French monarch put him at odds with the politics of the angry crowd. However, his contributions as a scientist had so far deflected the crowd's wrath.

Things were about to get worse. In November 1793, a chilling new howl was added to the fearful chorus: "*A bas les philosophes!*" (Down with the scientists!) The mob, it seemed, had decided that even scientists were part of the hated elite. The *sans culottes* were now looking for Lavoisier. He went into hiding but soon gave himself up to the authorities. He remained hopeful that his lack of direct involvement in politics and government, supplemented by his reputation as a world-renowned scientist, would free him. For seven months, Lavoisier languished in jail, doing what he could to defend himself against trumped-up charges.

He was tried on May 8, 1794. A fair trial was beyond hope, and a trip to the guillotine was assured even before the trial began. To the evidence presented by his friends regarding Lavoisier's good and important work, the trial judge, a large, dead-eyed man named Coffinal, replied dully, "The Republic needs neither scientists nor chemists; the course of justice cannot be delayed." Revolutionary French justice moved swiftly; Lavoisier and his head were soon separated in front of a cheering throng of rabble.

So ended the life of Antoine Lavoisier, the father of modern chemistry and the genius who unlocked the secret of oxidation. However, his work and reputation lived on. Forty years later, Lavoisier's chemical principles were so widely accepted that Michael Faraday described Lavoisier's oxidation theory to schoolchildren during the Royal Institution's Christmas Lecture.

OXIDATION

From a purely practical standpoint, fire and its uses have been well understood for a very long time. For millennia, people have known how to kindle it, tend it, and extinguish it. But from a theoretical standpoint, it was a mystery until Lavoisier arrived on the scene.

Phlogiston theory, the successor to the Four Element theory (all things are made from earth, air, water, and fire), was in vogue throughout most of the 18th century. Its proponents posited that all burning and flame are the result of drawing out an invisible flammable substance, phlogiston, that is contained within the fibers of things that are burnable. Phlogiston, went the theory, is the reason some things, such as wood and coal, burn, and other things, such as stone, do not.

The first person to isolate oxygen was Benjamin Franklin's friend and protégé Joseph Priestley. Priestley isolated oxygen in 1774 but had no real idea of what he had discovered. He began by studying mercuric oxide. Using a large convex mirror purchased from an Italian prince, he focused the sun's rays on a glob of red, toxic mercuric oxide inside a closed glass container. Paradoxically, when heated, mercuric oxide releases a gas that is anything but poisonous. Animals breathe it happily, and it causes flames to glow with incredible brightness. Priestley thought at first that the vapor in the jar was just exceptionally pure air.

In a preceding chapter, we saw how Priestley conducted several experiments with this mysterious vapor. He found that burning a candle in an enclosed space extinguished both the flame and any creature therein, yet growing plants in the same enclosed space restored life-sustaining conditions. Priestly concluded that he had somehow removed a specific substance, phlogiston, from the air, thereby making the air better at supporting burning and respiration. He called his discovery "dephlogisticated air." He reasoned that because it was able to support burning particularly well, it must have the ability to absorb great quantities of phlogiston, and so must be particularly devoid of it, or dephlogisticated.

A few years after Priestley, Lavoisier realized that the substance Priestley originally discovered wasn't simply a modified form of air but rather a separate component contained within a mixture of gases, and that this component, which he named oxygen, had some pretty special properties. In one of his best-known experiments, Lavoisier carefully ignited a piece of highly refined iron wire inside a container of purified oxygen. After burning the iron in oxygen, he precisely collected and measured what remained. He found that the burned material actually weighed a little more than the preburned material. That of course would be impossible under the phlogiston concept. Phlogiston, if it had really existed, would have been released and burned, making the burned object lighter, not heavier.

Lavoisier reasoned that fire and burning were not reactions in which something intangible (phlogiston) was *separating* from a burning object. Instead, he found that a separate element (oxygen) was *combining* with the burning object in a self-sustaining chemical reaction and that this reaction produced heat.

This realization changed everything. Scientists finally knew what fire is: a chemical reaction that always involves the combination of the burning material with oxygen. The discovery of oxygen and the role it played in fire and flame ended the old age of alchemy, of which the phlogiston concept was the last gasp, and dispelled old-fashioned ideas of the Four Elements.

– – – – – – – – – – –

The general equation that describes fire and combustion is:

Chemical + oxygen → chemical oxide + heat

Substances that burn (hydrocarbons such as wood, paraffin, and natural gas) are capable of combining with oxygen, converting to an oxide, and liberating heat. Substances that cannot combine with oxygen do not burn.

Here is a typical hydrocarbon-fueled fire equation, using propane, a hydrocarbon:

$$C_3H_8 + 5O_2 \rightarrow 3CO_2 + 4H_2O$$

To a practicing chemist, the level of description here might seem simplistic because the actual chemical reactions are indeed complex, involving chemical bonds, the transfer of electrons from one molecule to another, and so on. But for our purpose of opening a small window into the chemistry of fire, this simple explanation highlights the importance of oxygen and identifies the unchanging basic physical processes that produce fire.

Oxyhydrogen: Exploding Bubbles

www.Video5.ThePracticalPyromaniac.com

This project dramatically demonstrates a simple oxidation reaction, namely the oxidation of hydrogen to form water. Be advised, it is a very exothermic reaction!

In the previous chapter, we saw that hydrogen burns with a faint flame, giving off only a small amount of light while generating quite a bit of heat. We also saw that oxygen doesn't burn or ignite itself, but it sure does make other things burn faster and brighter.

You might already expect that when hydrogen is burned in oxygen, the vigor of the combustion reaction is greater. But you might be surprised at how much greater! When oxygen and hydrogen gas are mixed, the result is called oxyhydrogen. When oxyhydrogen is ignited, the reaction forms water and gives off a great deal of energy in so doing. The chemical formula is:

$$2H_2 + O_2 \rightarrow 2H_2O + heat$$

That formula may be chemically simple, but these exploding bubbles spectacularly display the results of the discoveries made by Cavendish, Priestley, and Lavoisier. This project builds on the successful completion of the Hydrogen Generator and the Pneumatic Trough detailed in chapter 7.

KEEPING SAFETY IN MIND

1. Follow the directions very carefully. This project is for responsible adults only. The exploding bubbles are soap

bubbles filled with a gaseous mixture of hydrogen and oxygen—oxyhydrogen—that are ignited. The sound made by the exploding bubbles is extraordinary, similar to or surpassing that made by a large firecracker or a short-barreled handgun. Eye and hearing protection are mandatory! Wear gloves to protect your hands as well.

2. The Exploding Bubbles project is to be demonstrated only in open air, on the surface of the Pneumatic Trough (see chapter 7 for Pneumatic Trough construction details). Never collect the bubbles and place them in a closed container.

3. Never ignite a bubble mass exceeding an area equivalent to a 2-inch-diameter circle. If more bubbles than that are present, pop the bubbles with a pin. Do not ignite big bubble masses!

4. As stated previously, always pour acid into water, never the other way around.

5. Vent the flasks frequently to prevent flammable gas buildup.

---

www.Project7.ThePracticalPyromaniac.com

Materials

General:

- Safety glasses
- Leather gloves
- Hearing protection (such as ear plugs or noise-blocking earmuffs)
- Measuring spoons
- Graduated cylinder
- Propane torch (for bending glass)
- Pliers (for bending glass)
- Scale, accurate to ½ gram or better
- Pin
- Long-handled lighter

Oxygen Generator:

- ❯ (2) 3-inchs length of 5-mm glass tubing with a 90-degree bend (See **diagram 2.7** for glass-bending instructions.)
- ❯ (2) 8-inch lengths of 5-mm glass tubing with 30-degree and 90-degree bends
- ❯ (1) 12-inch length of rubber tubing
- ❯ (2) 8-inch lengths of 4.8-mm rubber tubing
- ❯ (2) #6½ two-hole rubber stoppers
- ❯ (2) 250-ml flasks
- ❯ 50 ml 3-percent hydrogen peroxide (Available at drugstores Do not use concentrations of hydrogen peroxide higher than 3 percent.)
- ❯ 150 ml water
- ❯ 4 grams manganese dioxide (Obtain from a nonalkaline battery. See chapter 7 for instructions on how to extract manganese dioxide.)
- ❯ (2) clothespins or tube clamps
- ❯ Y-shaped tubing fitting

Hydrogen Generator:

- ❯ Hand file
- ❯ (2) post-1982 pennies with copper plating removed. (See the Hydrogen Generator section in chapter 7 for instructions for removing copper plating from pennies.)
- ❯ (1) 3-inch piece of 5-mm glass tubing bent into an L shape. (See **diagram 2.7** for glass-bending instructions.)
- ❯ (1) #6½ one-hole rubber stopper
- ❯ (1) 12-inch length of 4.8-mm rubber tubing
- ❯ 50 ml muriatic acid (available in hardware stores)
- ❯ (1) hose clamp or clothespin
- ❯ Y-shaped tubing fitting

Pneumatic Trough:

- ❯ Glass loaf pan, 1½-quart size
- ❯ Water to fill trough
- ❯ ½ teaspoon Dawn brand dishwashing liquid
- ❯ 18 inches of rubber tubing

BUILD THE OXYGEN GENERATOR SUBASSEMBLY

The oxygen generation subassembly allows you to generate a controlled amount of oxygen by adding small amounts of hydrogen peroxide to a flask containing manganese dioxide.

1. Begin by attaching the two 8-inch glass tubes to a two-hole stopper to make the hydrogen peroxide flask shown in **diagram 8.1.**

8.1 Oxygen Generator subassembly

2. Attaching the two 3-inch bent glass and rubber tubes to the two-hole stoppers in a second flask as shown at the top of **diagram 8.2.**
3. Fill the flask with the 8-inch tubes with a solution of 50 ml of hydrogen peroxide and 150 ml of water.
4. Add 4 grams of manganese dioxide to the flask with the 3-inch tubes.
5. Place the stoppers on the flasks and attach the open hose from the flask containing the manganese dioxide to one end of the Y-connector.

BUILD THE HYDROGEN GENERATOR SUBASSEMBLY

1. Expose the zinc in two post-1982 pennies by filing away as much of the copper plating as possible.
2. Insert a bent glass tube into the stopper.
3. Attach the 8-inch rubber tube to the top of the glass tube.

PREPARE THE PNEUMATIC TROUGH

1. Fill the glass loaf pan ²/₃ full of water and then add ½ teaspoon of Dawn dishwashing liquid. Stir gently.

CONNECT THE SYSTEM

1. Connect the rubber tubes from the Oxygen Generator, Hydrogen Generator, and Pneumatic Trough to the Y-fitting as shown in **diagram 8.2**.
2. Place hose clamps or clothespins on tubes at points A and B as shown in the diagram.

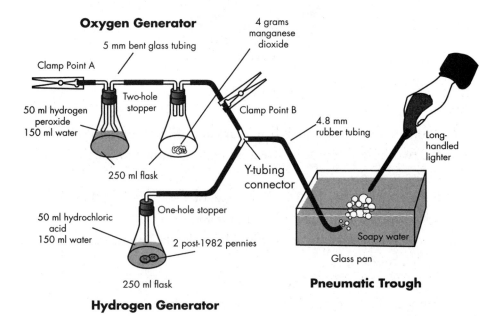

8.2 Exploding Bubbles assembly

GENERATE HYDROGEN

1. Pour 150 ml water into the Hydrogen Generator flask.
2. Carefully add 50 ml of normal hardware store strength (32 percent) muriatic acid.
3. Drop in two post-1982 zinc pennies that have had the copper plating removed. Stopper the flask. Connect the 8-inch rubber tubing to the glass tube in the stopper of the hydrogen generator. The other end terminates on an open end of the Y-connector, as shown in **diagram 8.2.**
4. Hydrogen gas will begin to bubble up and exit the flask through the tubing.

GENERATE OXYGEN

1. Remove the clamps from points A and B. As shown in **diagram 8.3,** pour a small quantity of the diluted hydrogen peroxide from Flask 1 into Flask 2. Oxygen gas will immediately be generated. Pour only small amounts (5 ml or less) at a time of the diluted hydrogen peroxide into Flask 2 to limit the rate at which the manganese dioxide catalyzes the hydrogen peroxide and produces oxygen. Set down Flask 1 and securely clamp it at clamp point A with a clothespin.

8.3 Generating oxygen

2. At this point, hydrogen is flowing from the hydrogen generator into the Y-connector and into the trough, as is oxygen from the oxygen generator. It will take a few moments for the oxygen gas to purge the air from the air hose. The initial bubbles will be a hydrogen/air mixture until the oxygen has cleared the air from the line.

MAKE YOUR BUBBLES OF OXYHYDROGEN

1. After the air is purged from the oxygen line, the bubbles being formed in the trough are filled with a powerful combination of hydrogen and oxygen. Wearing safety glasses, gloves, and hearing protection, ignite the individual, very small groups of bubbles with the long-handled lighter, as shown in **diagram 8.2**.

   The noise and energy released when the two elements react is tremendous and very entertaining. Popping these bubbles with the lighter is a lot like popping bubble wrap, but much more fun! Although water is formed in the reaction, it is not noticeable in the trough.

   The bubbles can be easily arranged in patterns on single or double bubbles in the trough and then ignited. This leads to several areas of experimentation which are left to the reader's imagination.

   Remember, never ignite a bubble mass exceeding an area the equivalent of a 2-inch-diameter circle. If more bubbles than that are present, pop the bubbles with a pin or your fingers. Wear safety glasses and other gear. Use caution and common sense at all times.

2. When you are done, remove all stoppers from flasks and vent the contents to the atmosphere.

3. Clean your equipment and dispose of the acid very carefully. Pour the discarded solutions down the laundry drain in your home; do not pour them into the storm sewer.

9

THE FIRE PISTON

When John Dalton, for years one of the leading lights at the Royal Institution, died in August 1844, the extent to which his hometown of Manchester, England, mourned was unprecedented. His black-draped mahogany coffin was placed in a darkened apartment and illuminated by artificial light in the city's great Town Hall. During the public wake, thousands filed past for a last look at the eminent scientist.

If Dalton's spirit was looking down from above, he no doubt would have found the building in which he lay in state to be a pleasing locale. With its classical Greek architecture and great marble columns topped by Ionic-style capitals, it was quite similar to the Royal Institution's building on Albemarle Street in London, where Dalton had lectured, listened, and passed many intellectually stimulating afternoons.

A funeral of such scale and expense had likely never been witnessed before outside of London. There were nearly 100 private carriages in the funeral procession, and 400 policemen were on duty, each with an emblem of mourning. The mile-long funeral train, estimated at 40,000 persons, included marchers, horsed riders, and carriages. The windows and the rooftops along the funeral route were lined with spectators waving a final good-bye to Manchester's beloved scientist.

It's hard to imagine such an outpouring of emotion for the passing of any modern scientist, but in those days it was the well-known scientists, not athletes or actors, who got the star treatment. Still, one might wonder just what Dalton did to deserve such adulation.

John Dalton's great contribution to science was to promulgate the basic concept that all things are made of indivisible particles called atoms. There are many types of atoms, said Dalton, each type having a different weight and each weight corresponding to a different element. Going further, he determined that when elements combine to form compounds, they do so not willy-nilly, in random proportions, but always in the same specific whole-number ratios. For example, water is composed of two parts hydrogen and one part oxygen. If a substance has other types of atoms or the ratios are different, then it's not water. That's obvious now, but it was a new concept in Dalton's time and is one of the foundations on which modern chemistry is built.

That alone would be enough to include Dalton in the pantheon of world-changing scientists, but he discovered even more. In 1801, he published a paper in which he declared, "All elastic fluids expand the same quantity by heat." That may not sound like a big statement, but in actuality it's a game changer, for Dalton figured out that the pressure of a gas (of a given mass and volume) is directly proportional to the gas's absolute temperature. Put another way, if a given quantity of a gas is placed in a container and you increase the gas's pressure by squeezing it into a smaller volume, then you increase its temperature as well. The converse is also true: decrease the pressure and the temperature drops.

That is a powerful idea, and it is the reason that fire-based things like internal combustion engines work the way they do. It's the reason German engineer Rudolf Diesel could invent his eponymous engine.

Dalton cast an immense shadow, and it's certain that Diesel was familiar with Dalton's work. But Diesel didn't come up with his ideas by simply reading Dalton's papers. He was also inspired in the way that Isaac Watts would suggest and advocate—first by a demonstration and lecture given by his teacher Carl von Linde at the college where he worked, then by personal observation, and finally by his own hands-on experimentation.

THE FIRE PISTON

Carl von Linde, considered the father of the modern refrigerator, had just returned to his home in Germany from a lecture tour that took him to Malaysia, among other places. This being the mid 1870s, the voyage had taken him months. Von Linde had seen and learned much during his excursion to Southeast Asia, and as a faculty member of the prestigious Munich Technical University, he was obligated to present the results and findings of his trip to students and faculty.

During his lecture, the fatigued Herr Doktor felt the need for a nicotine hit. He paused and withdrew from his pocket a small wooden cylinder and plunger that he likely called *"ein Feuerkoben."* The small device was a present from the people he had met on Penang Island in the Strait of Malacca. The indigenous people of the region used it to start fires. A person experienced in the use of the Feuerkoben, or fire piston, could reliably provide hot glowing embers anytime they were needed, even in the humid conditions of the rain forest.

At the lectern, Linde slapped the plunger down, and the tinder inside ignited. He plucked out a glowing ember and lit his cigarette with it. It was a neat gesture; to the audience, it looked like he had produced fire from nothing at all—no match, no flint. The fire had magically appeared from the bottom of an empty, hollowed-out tube.

The concept was not lost on audience member Rudolf Diesel. One of Professor von Linde's most promising students, Diesel had been experimenting with the recently invented internal combustion engine and was growing frustrated with the inherent low efficiency of the spark-ignition cycle engine. When von Linde lit that cigarette, a question sparked in Diesel's mind: "Could the same thermodynamic process that ignited the tinder in the bottom of the fire piston also ignite fuel in an internal combustion engine?" If so, perhaps here was a way to significantly improve the efficiency of this type of engine. As history proved, it was indeed.

Unlike typical gasoline engines, the now ubiquitous Diesel engine has no spark plug or carburetor. Instead, the Diesel engine works by compressing fuel under very high pressure. When the fuel/air mixture in the cylinder is compressed, it also gets very hot. In fact, it quickly exceeds the flash temperature of the fuel and ignites. The compressed gas expands violently upon ignition and pushes the compressing piston away with enough force to easily turn a drivetrain.

Scientists had known from Dalton's work that compressing a gas in a closed, insulated space causes it to get hot. At the turn of the 19th century, Dalton and French scientist Joseph Gay-Lussac independently conducted a series of experiments that proved that the temperature of a fixed mass and volume of gas are directly proportional to the gas's pressure. But it fell to Rudolf Diesel to figure out how to use this knowledge to make a high-efficiency engine that could work with no need for a spark. Diesel published a paper in 1893 outlining his ideas for a spark-free compression-combustion engine. In 1897, he built the first working compression-ignition internal combustion engine, and a little over 100 years later Diesel's engine is under the hood of millions of vehicles.

Was von Linde's fire piston the true antecedent of the modern Mack Truck engine? Accounts vary, but one thing is certain: the fire piston is not only fun to make and use, it's scientifically interesting and historically significant. Here's how to make your own.

- - - - - - - - - - -

Making a Fire Piston

www.Project8.ThePracticalPyromaniac.com

Materials

- Lathe or table saw
- (1) clear polycarbonate or acrylic rod, ½-inch-diameter × 12 inches long (the piston)
- Epoxy glue
- (1) clear polycarbonate or acrylic rod, ½-inch-diameter × 1¼ inches long (the plug)
- (1) clear polycarbonate or acrylic tube (the cylinder), ½-inch ID (inside diameter) × ⅝-inch OD (outside diameter), 9½ inches long
- Medium grit sandpaper
- ½-13 threading die
- (1) 1⅝-inch-diameter ball knob, with a hole threaded with a ½ inch-13 female thread x ⅝-inch depth (Note: A PVC tee fitting, ½ × ½ × ½-inch, may be substituted for the ball knob.)
- (1) pad of steel wool
- (2) ¼-inch ID × ⁷/₁₆-inch OD x ³/₃₂-inch rubber O-rings
- Drill and ⅛-inch drill bit
- Petroleum jelly

117

9.1 Cutting O-ring groove

1. Using the lathe or a table saw, cut the groove for the O-ring about ¼ inch from the end of the 12-inch rod. The depth of the groove should be just slightly less than the diameter of the O-ring. If the groove is deeper, the O-ring won't seal against the tube properly. If the groove is too shallow, you won't be able to insert the rod into the tube.

   The best way to cut the groove is with a lathe, but if you don't have one you can improvise by using a table saw. This method requires a bit of trial and error. Raise the blade so the height of saw blade protruding over the table equals the diameter of your O-ring, minus a hundredth of an inch or two. Carefully spin the rod as it contacts the blade to make an even slot. You might not be successful on your initial tries, but the rod is long enough so you can cut off mistakes and try again.

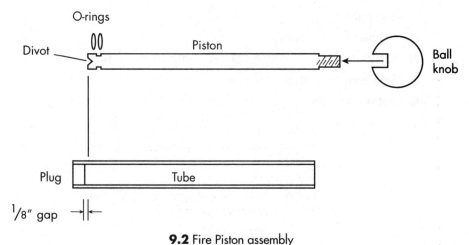

9.2 Fire Piston assembly

Refer to **diagram 9.2** to complete the following steps.

2. Using epoxy, glue the short rod into an open end of the tube. You may have to sand the plug slightly to fit into the tube. Do not over sand, as it is very important that the glue makes the end airtight. Rotate the plug in the tube to distribute the glue.

3. Use the threading die to cut a ¼-13 inch thread on the other end of the rod. Screw the ball knob onto the thread. Alternatively, you can glue the rod into the middle hole of a ½-inch PVC tee fitting for a handle. If you do this, you won't need a die.

4. Insert the rod in the tube and check the sliding fit. Use sandpaper and steel wool to make a close but free sliding fit between the rod and tube. Cut the rod to a length of about 9½ inches. Optimally, there should be about ⅛ inch of space separating the end of the rod from the plug when fully inserted.

5. Install the O-ring into the slot. Depending on the width of the cut you made and the shape of the O-ring you procured, one or two O-rings will fit in the slot. Sand the O-rings if necessary to obtain a smooth sliding seal.

6. Drill a ⅛-inch hole ("the divot") ³⁄₁₆ inches deep in the middle of one end of the rod. Your fire piston is now complete!

9.3 Finished Fire Piston parts

TIPS AND TROUBLESHOOTING

Test for air leaks. Smear petroleum jelly on the O-ring and carefully insert the rod into the tube, working the O-ring past the edge of the tube. If you've done everything correctly, the piston will smoothly and easily pop back up, nearly to the top, when you press down then release the knob. If you press on the piston and it just stays in the tube, then the fire piston won't work.

If this happens:

1. Check for leaks in the plug by spraying the end with soapy water, compressing the piston, and looking for bubbles.
2. Improve the sliding fit by adjusting the depth of the O-ring groove and repolishing the surface of the piston.

USING THE FIRE PISTON

1. Place a pinch of combustible material in the divot in the end of the rod. The best material is called charcloth. (See page 121 for instructions on making charcloth.)
2. Smear more petroleum jelly on the O-ring.
3. Carefully insert the rod into the piston, working the O-ring gently past the edge of the tube. Place the plug end of the fire piston on a hard surface. Quickly and firmly press down on the knob as shown in **diagram 9.4**. You'll see a bright flash in the bottom of the fire piston.
4. Carefully remove the piston from the tube and blow on the glowing charcloth in the divot, shown in **diagram 9.5**. You can now use the smoldering ember to start a larger fire.

9.4 Using the Fire Piston

9.5 Lit charcloth

How to Make Charcloth

What is charcloth? Basically, it's cotton cloth that's been roasted or pyrolized at high temperature in the absence of air. The wonderful thing about charcloth is that it is very easy to ignite with just a small spark. Charcloth doesn't burst into flame when ignited, but it does easily catch fire and smolder, making it just right for starting something else on fire, such as tinder or even a cigar.

Materials

- ◉ Charcoal grill and briquettes
- ◉ Long-handled lighter
- ◉ 10d nail
- ◉ Hammer
- ◉ Airtight metal can with cover, such as a hard candy tin
- ◉ 4 × 4-inch square of 100 percent cotton cloth cut into 1-inch squares
- ◉ Heatproof tongs

1. Light the charcoal briquettes in the grill and wait until they turn white.
2. Use the nail to punch a small hole in the top of the metal can.
3. Place the cotton squares in the tin and replace the top.
4. Place the container on the hot charcoal briquettes as shown in **diagram 9.6**. Almost immediately, the cloth inside will start to roast and white smoke will pour out of the hole. After several minutes, the smoke volume will decrease or stop, signaling that the charcloth is done.

9.6 Roasting charcloth

5. Using heatproof tongs, remove the tin and let it cool. Once the container has cooled, you can remove the top and take out the charcloth. The charcloth is ready when it is black, not brown, and fairly stiff but not completely brittle.

9.7 Charcloth

⚠ **10** ⚠

THE ARC LIGHT

Sir Humphry Davy
Abominated gravy.
He lived in the odium
Of having discovered sodium.

—E. Clerihew Bentley

Humphry Davy was an intellectually gifted child who excelled beyond his schoolmates in almost all subjects. By the time he was a teenager, he was a competent chemist who amazed his friends with interesting experiments. One such experiment was the creation of what he called "thunder powder," a pyrotechnic powder that provided a great flash and report.

Davy's father, a woodcarver in the southwestern English county of Cornwall, died when Davy was 16. Apprenticed to a surgeon, Davy seemed destined for a career in medicine. But it was chemistry, not medicine, that called Davy. A Cornwall scientist named Davies Gilbert knew of Davy's abilities in both medicine and chemistry and recommended him for a chemist's job at the new Pneumatic Institution for Inhalation Gas Therapy in Bristol in 1799. Thomas Beddoes, an influential doctor and tenacious fundraiser, had founded the research facility to study "the chemistry of airs."

The job was a good fit for Davy, combining his years of medical education with his intense interest in the burgeoning field of chemistry. At the Pneumatic Institution, he experimented with the therapeutic effect of gases on human ailments, hoping to cure such diseases as "palsy, dropsy, venereal disease, and scrofula" by the inhalation of various vapors and gases. In 1800, Davy published a paper on his discovery of nitrous oxide. Commonly known as happy gas or laughing gas, nitrous oxide is a chemical compound (N_2O) that has the ability to ameliorate pain. Although it took many years for nitrous oxide to become popular as an anesthetic, Davy's discovery was an important breakthrough in medicine and elevated his professional status significantly.

Besides his superior intellect, Davy also possessed excellent social skills and was very good-looking. Women, his biographers wrote, "swooned over him." With so much going for him, it was only a short time before he left Bristol for London, assuming the position of assistant lecturer at the Royal Institution in 1801.

One of the most popular social activities in Georgian London was attending lectures, and the most popular were those on scientific topics. If the lecturer was well known and possessed reasonably good stage presence, hundreds of people would turn out and were willing to pay high admission fees. Davy was a top celebrity of his day, filling the capacious auditorium of the Royal Institution with crowds eager to see the dynamic experimenter in action. His lectures were packed with vigorous chemical reactions, giant sparks, and the great carbon arc light featured in this chapter. However, Davy did not allow his lectures to devolve into mere pyrotechnical displays; he made sure that providing information about the scientific projects on which he was working remained foremost. And he was certainly working on a lot of projects.

As good a lecturer and fundraiser as he was, Davy was an even better scientist, an indefatigable experimenter and scientific pathfinder. In addition to discovering nitrous oxide, Davy's accomplishments include the initial isolation or discovery of more elements than any other person had discovered; his list includes barium, boron, calcium, magnesium, sodium, and potassium. Davy's work in electrochemistry is seminal to our understanding of chemistry today.

THE ARC LIGHT

Thomas Edison did not invent the first electric light. Although he likely deserves credit for the first incandescent electric lightbulb, more than 70 years before Edison's 1879 incandescent lamp patent, Humphrey Davy developed a technique for producing controlled light from electricity. He called his invention the arc light.

Like a candle flame, the arc light is hot and luminous. However, it is a different sort of fire; in many ways the hot arc is different from the chemical oxidation reactions we explore elsewhere in this book. But there are similarities as well, making the arc light experiment well worth pursuing in our attempt to fully understand the nature of fire by hands-on observation.

Davy's artificial electric light consisted of two carbon rods, made from wood charcoal, connected to the terminals of an enormous collection of voltaic cells. (In Davy's day, thousands of cells, similar to modern chemical batteries, had to be wired together in series to produce the voltage required to strike an arc between the carbon electrodes.) When Davy closed the switch connecting the battery to the electrodes, electricity jumped between the carbon tips. The resulting lucent dot of white heat glared so brightly that it was dangerous to look at for more than a split second.

While making an arc light isn't terribly complicated, the arc's underlying physical processes are indeed complex. Although normally a nonconductor, carbon will conduct electricity under certain circumstances. The graphite rods used in arc lights conduct electricity, albeit grudgingly, if enough electrical potential is applied. At high-voltage levels, the rod tips become white-hot, and carbon particles break away from the main body of the rod. Within the resulting particulate mist, small bits glow white with heat and jump across the spark gap between electrodes. This produces the incandescent arc of light known as an electric arc.

"The Dazzling Splendor," as Davy called it, was a tricky beast to control. After the initial sparks appeared between the electrode tips, Davy had to separate the carbons slightly and carefully to sustain the continuous bright arc of electricity. Once that was accomplished, he found the device could sustain the arc for long periods of time, although the carbon rods were consumed unevenly, making it difficult to keep the intensity of the light constant.

Davy's arc light was not economically practical until the cost of producing a 50-volt or so power supply became reasonable. This didn't occur until the mid-1870s, when Davy's protégé Michael Faraday invented the electrical dynamo. The dynamo made it possible to economically generate the high-voltage electricity needed to break down the carbon in the electrode tips.

Shortly after that, an American inventor named Charles Brush developed an arc light with an ingenious self-adjusting spark gap that solved the uneven light output problem that had made the arc light unacceptable for many applications. With those advances, arc lights became common, archetypically seen in searchlights, as well as in lighthouses, in street lights, on movie sets, and in movie projectors.

It took a lot of juice to run a searchlight. To maintain its arc, a 60-inch-diameter World War II–vintage carbon arc searchlight drew about 150 amps at 78 volts, which is equivalent to a 12,500-watt lightbulb. A lot of power, yes, but it could light up an airplane five miles away. Perhaps the largest carbon arc lamp ever made was the 80-inch-diameter monster searchlight that General Electric built at the turn of the 20th century. It lit the grounds of the 1904 St. Louis World's Fair with one billion candlepower!

Making a Davy Carbon Arc Light

www.Video6.ThePracticalPyromaniac.com

www.Project9.ThePracticalPyromaniac.com

Materials

- (2) carbon rods, each ½ inch long x approximately ¼ inch in diameter (The easiest way to obtain a pure carbon rod is to cut open a regular *non*alkaline AA, C, or D cell battery with a Dremel tool or hacksaw. Such batteries are usually labeled "heavy duty" or "non-alkaline." Cut off the top and carefully remove the carbon rod from the black, greasy packing that surrounds it. The packing material, manganese dioxide (which we used in the chapter 7 Oxygenizer project), will stain hands, clothes, and work surfaces, so wear rubber gloves and cover surfaces with newspaper.

- Cotton cloth, to clean electrodes
- Medium grit sandpaper

- (2) 1½-inch lengths of ¼-inch-diameter copper tubing
- (2) posable alligator clips (Easily made by soldering an alligator clip to a stout copper wire and securing the wire to the wood frame with (2) ½-inch-long #6 wood screws)
- Screwdriver
- (1) 2 × 4-inch wood piece, about 12 inches long (the frame)
- Heavy-duty on/off switch
- Miscellaneous wood screws for mounting posable clips, mounting the on/off switch, terminating nichrome wire, etc.
- 2 feet of 20- to 24-gauge nichrome wire
- (2) ceramic insulators (Electric fence insulators work well.)
- Nuts and bolts for mounting insulators to the 2 × 4 wood base
- Heavy-duty, flexible stranded lamp cord
- Wire stripper for removing the insulation from the ends of the lamp cord
- 12-volt power supply, such as a battery charger with ammeter or an 18-volt battery (for example, from a portable power drill)
- 100 percent UV protection glasses

Refer to **diagram 10.1** while completing the following steps.

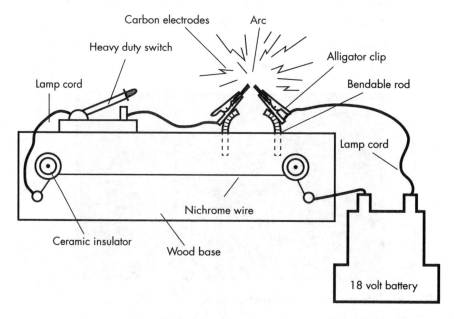

10.1 Arc Light assembly

1. The arc light's electrodes are made by carefully cleaning the carbon rods with a cloth to remove any manganese dioxide. Sand down the carbon rods until they fit snugly into the ¼-inch-diameter copper tubes. Crimp the copper tubes so they hold the rods snugly. Sand the protruding end of the electrode into a point.

2. Mount the posable clips to a wood frame and position the electrodes in the clips.

3. Mount the on/off switch on the top of the wooden base by fastening it with wood screws.

4. Mount the ceramic insulators approximately 10 inches apart as shown.

5. Wire the circuit using the lamp cord. Use the wire stripper to remove insulation from the ends of the wires. Electricity from the battery goes to the first electrode holder, through the carbon electrode, and across a very small spark gap to the second electrode. From there, the electricity goes through the main on/off switch and then across a length of nichrome wire before entering the opposite pole of the battery or battery charge.

6. Put on the safety glasses. Close the switch and carefully adjust the spark gap until a bright white light appears as shown in **diagram 10.2**.

7. Once a bright arc is struck and maintained, you can optimize the light output of the system by making the nichrome resistor wire longer or shorter.

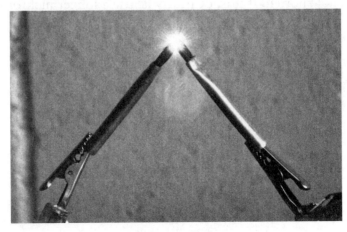

10.2 Arc of light

TIPS AND TROUBLESHOOTING

1. Every homemade arc light is a bit different. Make adjustments as necessary.
2. The spacing of the electrode gap is critical. Take your time adjusting it to obtain the best arc light. Too much or too little contact will result in no arc light.
3. Your battery will be damaged if the circuit is run without adequate load. The nichrome wire provides just enough resistance to prevent the battery from shorting. You will have to adjust the length of the nichrome wire for best performance by trial and error. If the wire is too short, it will quickly burn up. If the wire is too long, the arc light will be dim.

KEEPING SAFETY IN MIND

1. The arc light produces ultraviolet radiation and is very bright. Cover exposed skin and wear 100 percent UV eye protection. Never look directly at the light.
2. The nichrome wire and copper electrode holders get extremely hot. Be very careful to avoid touching the hot wire!
3. This arc light is a demonstration device only and should only be operated intermittently and briefly. Running the arc lamp too long can damage your battery or battery charger. If using a battery charger, check its ammeter to make sure the circuit is not shorted. If it is, or nearly is, use a longer nichrome resistor wire.

11

FIREPROOF CLOTH AND COLD FIRE

Al-Malik al-Zāhir Rukn al-Dīn Baibars al-Bunduqdārī, who is better known by the easier-to-remember name Baybars, was one of the more intriguing characters of the Crusader era. He reigned as sultan of Egypt from 1260 to 1277 C.E. and was, depending on the opinion of the author researching him, either one of the smartest or one of the most ruthless men who ever lived. Certainly he had to be one or the other, or perhaps both, for his ascent to power was one of the speediest in recorded history: he rose from slave to supreme leader of one of the largest empires on the planet in less than 30 years.

Baybars, known as the Lion of Egypt, was apparently fair-skinned, quite tall, blind in one eye, and possessed enormous political acumen. He was a man of scientific accomplishment as well, as much evidence points to the conclusion that Baybars was involved in the invention of fireproof fabrics.

According to historians, Baybars was born around 1220 C.E. in Crimea. Captured as a child by slave raiders, he was eventually recruited to become a member of the Mamelukes, an elite team of slave caste warriors allowed to carry weapons and even own other slaves. The Mamelukes were a powerful political force in Egypt, and within a short time Baybars worked his way up to be the group's commander.

While Baybars was managing his rise through the Mameluke ranks, history-shaking events were occurring a thousand miles to the east in Mongolia. In 1251 Genghis Khan's grandson, Prince Mongke, became the supreme leader, or Great Khan, of the Mongol Empire. One of his most significant acts was appointing his nephew Hulagu to lead the great Mongol horde of the southwestern steppes of central Asia, called the Ilkhanate.

Mongke had a specific mission for Hulagu, namely leading the vast army of Ilkhanate horsemen west to conquer and subjugate the Islamic empires of the Middle East. Obviously, such a plan did not sit well with the Egyptian leader, Sultan Qutuz, who was also the leader of a Mameluke faction that was a rival to Baybars's group.

Military and political skirmishes between Qutuz's and Hulagu's armies played out at intervals until things came to a head in 1260. Hulagu sent two envoys to Sultan Qutuz in Cairo with a blunt message demanding surrender. It read in part, "To cowardly Qutuz the Mameluke: Consider what happened to those nations who refused to submit to us. We have conquered vast areas, massacring all the people. Where will you go? What road will you use to escape us? Our horses are swift, our arrows sharp, our swords like thunderbolts, our hearts as hard as the mountains, our soldiers as numerous as the sand."

In response, the angry Qutuz had the messengers sliced in half and set their heads up on the gates of Cairo. The results of this act were predictable; the fight was on. Qutuz marched out to engage the Mongols. He was soon joined by Baybars's Mamelukes, who decided that the goal of stopping the invading force of Mongols swooping into Syria and Egypt from the east outweighed dislike for Qutuz.

The turning point of the war took place near a small desert community called Goliath's Well (Ain Jalut) near present-day Nazareth. Baybars led the combined Arab army in a momentous pitched battle. By executing a series of feints, diversions, and flanking attacks, Baybars's army cut down the Mongols' main column of horsemen, sending them fleeing westward in a disorderly retreat and effectively ending the Mongol threat to the Islamic kingdoms of the Middle East. Never again would the fleet, mounted archers from the steppes threaten the cities rimming the Mediterranean.

Upon returning to Egypt, Baybars, conniver that he was, ordered the assassination of Qutuz. He deftly maneuvered himself into power, becoming the sultan and reigning for the next 17 years, until he drank poisoned wine and died.

An interesting aspect of Ain Jalut is that it is perhaps the earliest battle in which gunpowder weapons were used. Baybars's soldiers were among the first to possess handheld firearms. Much different from the firearms of today, they were extremely primitive, inaccurate, and hard to use. It is said that they were nearly as dangerous to the people who used them as they were to the people they were aimed at. Nonetheless, it is highly likely they were of great value because the noise and smoke they produced created fear and confusion among the Mongolian horsemen.

Given the rough state of firearm and gunpowder technology, the more farsighted soldiers in Baybars's army started thinking of ways to improve the odds of success. Their solution? Flame-resistant clothing that would protect soldiers and their horses from the fire and smoke produced by the powerful, unpredictable weapons. The *St. Petersburg Manuscript*, an ancient handwritten book, describes the fire-resistant armor Baybars's Mameluke horsemen wore during the battle of Ain Jalut:

This was the practice of the time of Hulagu, when the people of Egypt used this trick to defeat the Mongols: Horses (of the enemy) dare not face fire and the horse will run away with its rider. So choose a number of knights and furnish their lances from both ends with gunpowder.

The knight will wear a garment with its front face made of specially made, fireproof, coated woolen cloth called "balas." Inside, it is lined with balls of linen fiber held in place with metal wires. The horse is also draped with coated woolen cloth. Also, the knight's hands will be smeared with dissolved talc so that he is not burnt by fire.

- - - - - - - - - -

Making Cloth Fireproof

Few things short of asbestos pants are absolutely fireproof. But just as 12th century Mameluke soldiers could figure out a way to make cloth resist fire, it is possible for 21st-century garage inventors to do the same.

www.Project10.ThePracticalPyromaniac.com

Materials

- Scale
- 8 ounces alum (Potassium aluminum sulfate.) (This chemical can be purchased in small quantities in the spice aisle of most grocery stores. Larger quantities are available in stores selling canning and pickling supplies and in many shops online.)
- Plastic spoon
- 8 ounces hot water
- Quart-sized mixing bowl
- 9 x 9-inch glass pan
- Cotton cloth, approximately 6 x 6 inches
- Drying rack
- Long-handled lighter or fireplace match

1. Weigh out 8 ounces of alum. Stirring rapidly with the plastic spoon, dissolve the alum in 8 ounces of hot water in the mixing bowl.

2. Pour the alum solution into the glass pan. Dip the cloth into the solution, saturating it completely.

3. Let the excess liquid drain off, and then hang the cloth on the drying rack to dry.

4. When completely dry, test a small piece of the fabric by holding a flame to it. If the fabric burns, increase the concentration of alum in the solution and repeat.

The chemistry behind how alum works is somewhat complex. Researchers have found that as alum decomposes in heat, it releases noncombustible gases that dilute the volatile ones, making burning difficult if not impossible. Basically, the alum treatment reduces the rate at which flames propagate through the fabric.

PROTECTING AGAINST FIRE

Joseph Louis Gay-Lussac, one of the most important French chemists of the 19th century and the father of modern fireproofing, was just 15 years old when Antoine Lavoisier met his demise on the guillotine. In the spring of 1794, the story was front page news all over France. Studying Latin and religion in a provincial school south of Paris, Gay-Lussac may have been too young to understand Lavoisier's contributions to science. But he did know that his father, like Lavoisier, was an officer in the employ of Louis XVI. The execution of men not unlike his father in the Place de la Revolution must have been most unsettling. Happily for him and his family, within a few years the storm had passed, the Reign of Terror ended, and life in France returned to relative normalcy.

Young Gay-Lussac moved to Paris and began to study the sciences with a friend and coworker of the late Lavoisier, the eminent scientist Claude-Louis Berthollet. Gay-Lussac's interest in chemistry grew as he worked from Lavoisier's seminal textbook on chemistry, *Traité Élémentaire de Chimie* (Elementary Treatise on Chemistry). The young man showed great aptitude and dexterity for chemistry, and he learned quickly under Bethollet's tutelage.

Like Dalton, Cavendish, Priestley, and Lavoisier, Gay-Lussac was a skillful experimenter. His contributions to physics and chemistry were tremendous. He is perhaps best remembered for his work on the nature of gases. He investigated the way they behave at different temperatures and pressures and was the first to find that the temperature of a gas varies proportionally with its volume.

Besides this, Gay-Lussac was one of the leading experimenters in the field of fireproofing. Working in his Paris laboratory, he performed and documented many experiments that formed the basis of the art and science of making flammable things less so, and in so doing making the world a bit safer.

Sulfur, or brimstone as it was once called, is the element most often associated with starting fires. Boron, on the other hand, is the element associated with stopping them. Boron is a good fire retardant because it chemically transforms the materials it treats, inhibiting the spread of flame and promoting the formation of a protective layer of char that acts as a fire barrier. It's often used to make materials fire resistant, notably paper and fabric.

For centuries prior to 1800, boron was widely used in compounds such as borax and boric acid, but the element itself had never been isolated. In the early 19th century, Gay-Lussac and another scientist, Humphry Davy, one of the lead scientists at the Royal Institution in London, were locked in a heated competition to isolate this element and lay claim to being its discoverer.

Gay-Lussac and Davy were similar in a number of ways. They were both born in 1778; each man was a world-famous chemist, widely known for exceptional laboratory technique and meticulous observation; and each man nearly lost his eyesight as a result of a chemical explosion. In 1808, the two scientists were neck-and-neck in a tight race to isolate the unique element locked inside the compounds of boric acid and borax. To world-class physical scientists like Gay-Lussac and Davy, the drive to discover new elements was a powerful one; doing so meant fame and prestige. Isolating and naming an element with properties as useful as this one's was a particularly compelling goal.

By 1808, Davy had already discovered and named five elements: barium, calcium, strontium, sodium, and potassium. He isolated sodium and potassium by using the giant electric battery or "cell" at his disposal at the Royal Institution. He thought, quite reasonably, why not try the same approach with boric acid? He made significant progress, electrochemically isolating small quantities of a previously unknown substance. However, he was unable to prove categorically that the substance he obtained was in fact a new element.

After hearing of Davy's work with boric acid, Gay-Lussac redoubled his efforts, not wanting to fall behind the Englishman in chemical exploration. Abandoning caution, Gay-Lussac and his assistant Louis Thenard (who went on to discover hydrogen peroxide, the compound we used in the Oxygenizer project in chapter 7) adopted a dangerous laboratory technique involving highly reactive pure potassium metal. Gay-Lussac and Thenard managed to isolate a substance they called "bore."

Gay-Lussac and Thenard were able to verify that what they had found was indeed a new element and therefore were able to publish their findings: "These experiments prove conclusively that this body which we now propose to call bore, is of a definite nature, which can be placed beside carbon, phosphorous, and sulfur." So, who was the first to discover boron, Davy or Gay-Lussac? It depends on how you look at it. Both have their backers, and to many historians, the outcome is too close to call.

In 1821, Gay-Lussac began experimenting with boron-based compounds to render them fire-resistant. "Wool and silk fabrics," he wrote "are not very combustible, while fabrics of hemp, flax and cotton take fire easily and burn with great rapidity. Consequently, it is those fabrics that it is more important to make incombustible."

He began by coating fabrics with metal salts and soon found that the salts of boron produced the best results. His success in finding a fireproofing chemical that wouldn't affect the color of cloth or turn it poisonous was seminal, allowing other scientists to build upon his work to make fire-resistant materials for use in theaters and other public spaces.

- - - - - - - - - -

Making Paper Fire-Resistant

This variation of Gay-Lussac's method of fireproofing paper and fabric appeared in the *Quarterly Journal of the National Fire Protection Association* in the 1920s and is still of considerable value. "Boronized" paper has been treated with a combination of boron salts that inhibits burning. Important papers can be treated in this fashion to give them a greater likelihood of surviving a fire.

> www.Project10.ThePracticalPyromaniac.com

Materials

- ⦿ 3$\frac{1}{3}$ ounces boric acid powder (Although boric acid is used as roach killer, it is relatively safe to handle and is often available in hardware stores or online.)
- ⦿ 4 ounces borax
- ⦿ ½ gallon hot water
- ⦿ Large mixing bowl
- ⦿ Spoon
- ⦿ Shallow tray (large enough to soak the paper)
- ⦿ Cotton paper (also known as rag paper, available at stationery stores)
- ⦿ Long-handled lighter or fireplace matches

1. Mix the boric acid powder and borax in ½ gallon of hot water in the mixing bowl. Stir with the spoon until the chemicals are completely dissolved.
2. Pour the solution into the shallow tray. Carefully place a sheet of cotton paper in the tray, saturating it thoroughly.

Remove the paper, allowing the excess solution to drain back into the tray. Hang the paper to dry.

3. When dry, test a small piece of the paper by holding a lighter or match to it. The paper will turn black but not ignite and burn.

Gay-Lussac found that fire does not occur if air can be prevented from reaching the surface of organic materials by chemically coating the materials' fibers. Silicates and borates are well suited to the task of rendering organic fibers flame-resistant. The normally flammable material calcinates—that is, turns to char or black powder—but does not actually go up in a flame.

Fireproofing has come a long way since the days of Gay-Lussac. Still, some places, such as theaters and auditoriums, require more protection than even the most modern fireproofing can provide.

— — — — — — — — — —

Psst! Avast there! It be too late to alter course, mateys. And there be plundering pirates lurkin' in ev'ry cove, waitin' to board.

> —The talking skull and crossbones
> at the entrance to the Pirates
> of the Caribbean attraction at
> Disneyland

In March 1967, the Pirates of the Caribbean attraction opened for visitors at Disneyland in Anaheim, California. In the last attraction that Walt Disney personally had a hand in designing, guided vehicles travel through carefully lit scenes that contain animation, sounds, music, and other special effects creating a pirate raid on an unfortunate Caribbean island town.

"Pirates" is well known among stagecraft buffs for its use of extravagant, realistic faux fire. At the tale's climax, the village of Isla Tesoro is set afire by drunken Animatronic pirates led by the fearsome Hector Barbossa; the entire room is filled with a red-orange glow mimicking a five-alarm fire. With so many people moving through the exhibit, the flames could not be real. Instead they are created out of fabric, stage lighting, and moving air.

From the time of the ancient Greek playwrights, actors and directors have depended on technical experts to create the sets, lighting, and effects necessary to realize the artistic vision of the play. Plots often require the use of fire; for example, staging Mozart's opera *Don Giovanni* requires everything from a cozy Franklin stove or Rumford fireplace scene to giant flames shooting from the gates of hell. It's the job of stagecraft experts to bring such images safely to the stage.

Exploring fire effects made from silk, air, and lights is an accessible entrance into the world of theatrical stagecraft. This project involves combining simple electronics with some mechanical know-how to create amazingly realistic faux fire.

Cold Fire

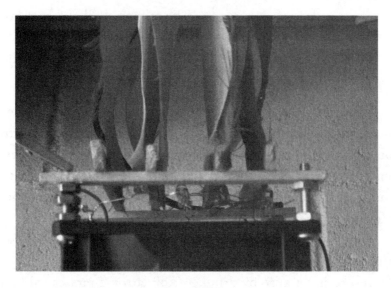

11.1 Cold Fire

To simulate a campfire, this project uses 12-volt electronics, inexpensive colored LEDs, and lightweight, brightly colored cloth.

140

The voltages involved are low and nonshocking, and the LEDs emit very little heat. That means that experimenting with different fire shapes, LED arrangements, and geometries is encouraged, although common sense dictates that you watch out for moving fan blades. You can use these instructions as a departure point for your own, perhaps more elaborate, faux fire projects. The project can be completed in an afternoon or two.

www.Project10.ThePracticalPyromaniac.com

MATERIALS

- (4) #8 machine screws and nuts
- Screwdriver
- (1) 2-foot length of $\frac{1}{4} \times \frac{1}{8}$-inch wood strip
- Fast-drying glue
- 12-volt fan (Several types of fans will work, including the common computer muffin fan and the centrifugal or "squirrel cage" fan. In general, the more air the fan moves, the larger and more realistic the flames will look. An 18 CFM fan will simulate small fires, and a 60 CFM fan will simulate larger ones.)
- (4) red, yellow, or orange ultrabright light-emitting diodes or LEDs (Ultrabright LEDs provide 5,000 to 10,000 mcd [millicandela]. Four ultrabright LEDs are perfect for making a realistic-looking small fire.)
- 200-ohm resistor
- 22-gauge hook up wire
- Scissors
- Approximately 1 square foot each of red and yellow silk or silk-like fabric (The lighter and sheerer the fabric, the better it will work.)
- (1) piece of hardware cloth, 3 × 10 inches
- 12-volt battery or power supply
- Dry ice (optional)
- Wood logs (optional)

PREPARE THE FAN

1. Fabricate a lattice out of the #8 machine screws and wood strips, using fast-drying glue. **Diagram 11.2** shows one way to build the lattice. You may need to improvise based on the dimensions and characteristics of the fan you procure, but the concept is similar in most cases.

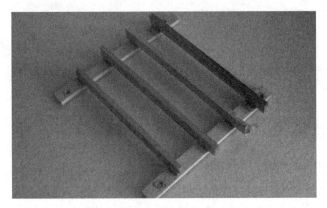

11.2 Cold Fire lattice

2. Once the lattice is complete and the glue has dried, you'll need to mount it on the outlet side of the fan. The fan casing has mounting holes at each corner. Insert the machine screws into the mounting holes. Mount the lattice securely to the fan casing using the machine screws.

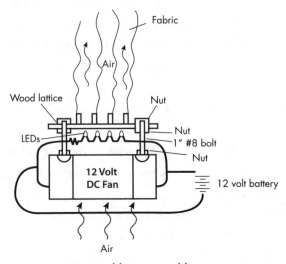

11.3 Cold Fire assembly

WIRE THE LEDS

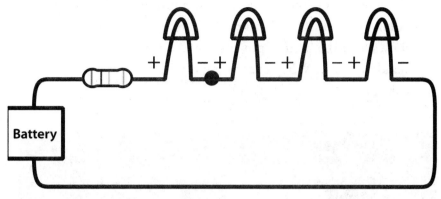

11.4 LED wiring

1. **Diagram 11.4** shows how the LEDs and the 200-ohm resistor are wired together in series. The register is necessary to limit the current, otherwise the LEDs will burn out almost instantly.
2. These instructions assume that the LEDs you are using have a forward voltage (usually denoted "VF" on specification sheets) of 2 to 4 volts, and a normal current draw of 20 to 30 mA. As you wire the LEDs together, take careful note of the two legs of the LEDs. The longer leg is the positive end. Wire the four LEDs positive to negative as shown in the diagram.
3. Attach the LED/resistor assembly to the lattice mounted to the fan housing.

CREATE THE SILK FLAMES

1. Use the scissors to cut the red and yellow fabric into flame shapes. The more powerful the fan you have (i.e., the higher the CFM rating), the longer the fabric pieces can be.
2. Glue or tape the fabric flames to the lattice.

ASSEMBLE YOUR COLD FIRE

1. Build a stand by bending the hardware cloth into a circle and taping it or folding over the edges to make it retain the circle

shape. Place the completed fire simulacrum upon it. This will allow better airflow to the fan and improve performance.

2. Hook up the LED and fan to the battery. **Diagram 11.3** shows how the battery powers the LEDs and the muffin fan simultaneously. When everything is wired correctly, the fan will spin and the LEDs will light.

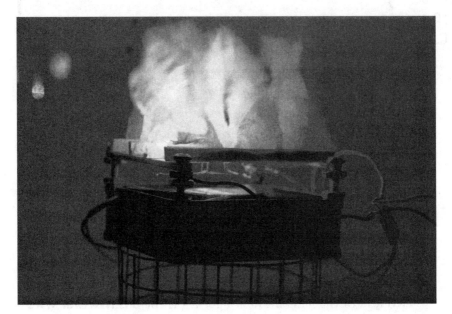

11.5 Cold Fire

TIPS AND TROUBLESHOOTING

When the fan turns and the LEDs light up, it is likely that your faux fire may not look very realistic initially. Don't worry; a little bit of tweaking will work wonders. Your fake fire may require the following adjustments:

1. The light emitted by the LEDs is very directional. Most of the light is aimed straight up and little goes out radially from the sides. Use this to your advantage by bending the LED wires to aim the light onto the fabric to produce the most realistic

lighting effects. Be certain to bend the LED leads slowly and gently; too much bending will cause them to break.

2. You can adjust the "flutter" of the flames by recutting the fabric into different shapes. Long, skinny shapes flutter more than shorter, flatter designs, although shapes that are too long will sag rather than wave. Once you determine the maximum length of fabric that your fan can handle, attach the longest shapes to the middle hoop and the shorter, flatter shapes to the end hoops.

3. You can improve the realism by placing the entire assembly on or near a bed of dry ice. The vapor from the dry ice will simulate smoke.

4. For the most realistic effect, hide the fan and the stand behind wood logs.

⚠ 12 ⚠

THE EXTINCTEUR

The people exploring the nature of fire at the turn of the 19th century were certainly an interesting lot. Some, like Lavoisier, Priestley, and Rumford, led lives full of color and adventure that prove the old adage that truth is stranger than fiction. Others, like George Manby, are memorable for other reasons.

Captain George Manby's friends would sometimes say he wasn't quite right in the head. When people say such things, they normally mean that the person is a bit eccentric or peculiar. But in the case of Manby, the inventor of the fire extinguisher, they were being quite literal.

In 1801, when Manby was a young man living in Wales, a man named Pogson shot Manby in the back of the head. Manby asserted that Pogson crept up behind him one night and shot him as he was getting into a boat. The Pogsonian version of the story was that Manby was shot while running away from a duel. Given the lack of evidence and the great amount of time that has passed, it is impossible to determine which, if either, story is true. What is certain is that Manby was wearing a woolen cap when he was shot.

When Manby was shot, the pistol ball pushed some of the felt fibers into his cranium. For a while, the wound did not seem life threatening, and Manby seemed satisfied with going through life hat in head, so to speak. However, the fabric's eventual decay inside his head led to an infection that necessitated the removal of the offending fibers, bullet fragments, and, unfortunately, parts of his brain.

It was, according to Manby, a terribly difficult and gruesome operation, but he pulled through. While Manby lived to be 89 years old and mostly in good physical health, he was never the same psychologically. He came to view the world in an odd sort of way and possessed an exceptionally strange view of his place in the world.

His correspondents and business contacts considered him a clever but unbalanced inventor at best and a presumptuous, self-aggrandizing nuisance at worst. His friends and acquaintances concluded that he was a bright fellow who suffered from delusions of grandeur—a difficult egomaniac who nonetheless had much to offer the world.

For all his quirks, Manby was an inventor of uncommon skill and was even elected a fellow of Britain's Royal Society. His contributions came not from his experimental technique or world-changing conceptual breakthroughs, but instead from his ability to recognize a problem and tenaciously attack it until solved. His first invention of note was a system for rescuing shipwrecked sailors. The system was basically a cannon-fired harpoon trailing a rope. When a boat foundered offshore, a harpoon was launched from shore at the sinking boat. Once a rope extended from shore to ship, a rescue craft could be launched. The system saved many lives. Manby's other invention was even more significant, however.

In October 1834, in London, Manby heard cries of alarm. He grabbed his coat and hat and ran outside. Everyone around him was rushing toward the buildings of Parliament. When Manby arrived, he saw big flames enveloping the House of Lords and the House of Commons and moving like a flaming curtain toward Westminster Hall. Valiant efforts were made to fight the fire, but the roofs of several of the old buildings were blazing and soon came crashing down, trapping firefighters in the burning rubble. Many were dead, and the great center of British government was in ashes.

Fires unfortunately were common in 19th-century London, and "fire-watching" was a popular form of entertainment. Large crowds would gather at even mundane fires, so a conflagration involving the seat of British government quickly drew thousands of unhelpful, obstructive spectators. The unruly crowds grew so big that three regiments of soldiers were called in to keep them from interfering with the work of the firemen. Manby himself narrowly escaped being trampled and run over by fire wagons.

> I reached for an iron railing, to which I clung until a Good Samaritan came to my aid. . . . He put me in a cab before I could get his name. As the cab drove off I caught a glimpse through the great doors of Westminster Hall. The fierce light of the fire reflected through the windows and exhibited a splendor awfully grand.

Westminster Palace survived, but Manby was shaken by his experience. He must have wondered how much better the outcome would have been had the Parliament buildings been equipped with the fire extinguisher he had developed a few years earlier.

The ability to fight fires, large or small, in London during the time of the Georgian kings was woeful. Fire departments were owned by fire insurance companies, whose main objective was to reduce property claims on insured buildings, not save lives. Typically fire departments responded only to fires on property owned by insurance company customers. The firemen themselves were underpaid, inadequately trained, and poorly equipped. Tools were limited to buckets of water or, in those lucky situations where rivers or lakes were nearby, crude pumps.

As a young man, Manby witnessed a large fire in Edinburgh, Scotland. Seeing the tragic destruction, he began to think about how fires could be fought more effectively. Years later, after much cogitation and experimentation, he came up with an idea. Manby's great innovation was to strategically place portable firefighting devices in the field so firemen could extinguish small fires quickly before they grew large. In a pamphlet describing his fire-extinguishing device, Manby wrote, "I have had the fact confirmed to me that fires generally become destructive from the long period of time that unavoidably occurs between the discovery (of the fire) and the assembling of the firemen, the arrival of the engines, the procuring of water and getting the engines into full action."

The original Manbian fire extinguisher was a sealed pot, nearly three feet long and large enough to store four gallons of a firefighting liquid that Manby termed "anti-phlogistic fluid." The fluid was sealed, under pressure, in a copper vessel. When the valve on top was turned, a powerful jet of fire-extinguishing fluid rushed out through a nozzle. The devices were placed on two-wheeled carts, designed to be pushed quickly to the site of the fire, before the blaze became uncontrollable. The "Extincteur," as Manby called it, was light enough for a man to lift and carry for short distances.

Manby's "anti-phlogistic" fluid was a solution of potassium carbonate and water. According to Manby, such a mixture was nearly 40 times more effective than plain water in extinguishing blazes. Today, modern "BC" type fire extinguishers use a chemical very similar to Manby's breakthrough: potassium bicarbonate. When sprayed on a fire, the chemical decomposes, releasing carbon dioxide and suppressing the oxygen that feeds the flames. Further, Manby wrote of the presence of a nonflammable residue on the extinguished material, which he said greatly reduced the likelihood of re-ignition.

In the late 1820s, Manby wrote a letter to Michael Faraday asking whether the Royal Institution, given its history of exploring things related to the science of fire, would be interested in a lecture and demonstration of his inventions. Faraday promptly wrote back, welcoming Manby to lecture at the Royal Institution. It was a prodigious moment for Manby, who wrote to a close friend that "Mr. Faraday, whose encouraging kindness gives me confidence and . . . [the Royal Institution's resources] will be at my service to make my preparation."

The lecture was well received by Royal Institution members, and Manby felt encouraged to continue developing his many projects. The fire extinguisher was a certainly good idea, and a number of influential people, including high-ranking military officers, were impressed. But Manby was not much of a businessman and was never able to realize the business potential of his invention. Still, he is remembered in history books as the inventor of what is likely the first fire extinguisher.

Portable fire extinguishers have come a long way since the Extincteur. Many different types of fire extinguishers are available, and all modern-day fire extinguishers are classified with a letter—A, B, C, or D—that shows which type of fires they are safe to use on and how effective they are. These letter designations are very important, as fighting a fire with the wrong type of fire extinguisher can be a big mistake.

Building a homemade fire extinguisher is simple and practical. Below are directions for building a "soda-acid" extinguisher, which was invented shortly after Manby's earliest pressurized models.

- - - - - - - - - -

Making a Mini-Extincteur

Early soda-acid extinguishers worked by breaking a bottle full of acid into a solution of water and sodium bicarbonate. A vigorous chemical reaction between the acid and the soda water produced prodigious amounts of carbon dioxide gas, which in turn shot the water out of an attached hose and onto the fire.

This fire extinguisher project works in the same way. We'll use sodium bicarbonate (better known as baking soda) and acetic acid (in the form of vinegar) to create copious amounts of frothing CO_2 that pressurizes a plastic container and allows us to direct a stream of water onto a blaze. Indeed, if you are making hydrogen, exploding bubbles, or undertaking any project involving the production of fire, then having this extinguisher handy is a wise idea.

Materials

- ◉ Scissors
- ◉ Newspaper
- ◉ 13-mm-diameter test tube or similar-sized rod
- ◉ Rubber cement
- ◉ Scale, accurate to 1 gram or better
- ◉ 18 grams baking soda
- ◉ 50 ml graduated cylinder
- ◉ 275 ml white vinegar (5 percent acetic acid)
- ◉ (1) empty and clean .5-liter (about 16 ounces) disposable plastic water bottle with "sports cap" (a spout you can open and close)

1. Use scissors to cut a rectangle of newspaper 5 inches by 3 inches. We'll use the test tube as a form to make a paper tube. Wrap the paper around the test tube and glue the edges and bottom with a very thin layer of rubber cement. Remove the test tube and let the paper tube dry.
2. Use the scale to weigh out 18 grams of baking soda. Carefully pour the baking soda into the paper tube.
3. Use the graduated cylinder to measure 275 milliliters of vinegar and pour the vinegar into the plastic water bottle.
4. Place the paper tube filled with baking soda in the neck of the bottle, making sure the paper does not touch the vinegar. Fold the top edges of the paper tube over the top of the bottleneck. Screw down the sports cap so that the cap holds the paper tube well above the surface of the vinegar as shown in **diagram 12.1**. Store the bottle carefully until you are ready to use it.

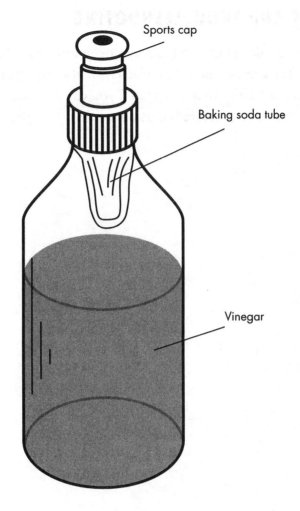

Sports cap

Baking soda tube

Vinegar

12.1 Mini-Extincteur

USING THE EXTINCTEUR

1. Make sure the sports cap is in the sealed (pushed down) position.
2. Shake the bottle vigorously, keeping the closed sports cap pointed in the direction of the fire. The bottle contents will react, generating carbon dioxide gas. Shortly after shaking, the pressure in the bottle will force open the sports cap and eject a stream of soda water onto the fire.

TIPS AND TROUBLESHOOTING

If the cap does not automatically open after you shake the bottle, open the cap manually. Conversely, if the cap opens too quickly, causing a feeble stream of water, you can manually hold it closed until adequate pressure builds up inside the bottle.

⚠ **13** ⚠

THE PHOTOMETER

In 1815, Humphry Davy received a letter from a colleague in the coal mining region of England, asking him if he could investigate the dangerous conditions that methane gas caused in mines. The gas, which miners called "fire damp," often filled the mines and was all too frequently ignited by the candles miners used for light. Many miners died in the resulting explosions.

Davy invented the coal miner's safety lamp in 1815, which used a permeable metal screen that supported combustion but did not allow the lamp flame to ignite vapors outside the lamp housing. While working on the lamp, he discovered some interesting things about the brightness of flames. One of Davy's more surprising finds was that the more soot a flame produces, the brighter it shines. Experimenting with a wire screen held over a flame, Davy found that by raising and lowering the screen he could adjust the amount of "solid matter" (now known to be carbon particles) undergoing combustion. "This principle of the increase of the brilliancy of the flame by the production and ignition of solid matter appears to admit of many applications," wrote Davy. "It offers means of increasing the light of burning substances. . . . Whenever a flame is remarkably brilliant and dense it may always be concluded that some solid matter is produced in it, and when a flame is extremely feeble and transparent, it may be inferred that no solid matter is formed."

As we discussed in the earlier chapter on candles, only a bit of a candle's light (the blue part at the bottom of the flame) comes from the chemical reaction between the fuel and oxygen. Instead, most of the light is produced when the unburned carbon particles, called soot, are heated to incandescence at the hot edges of the flame. Hotter soot appears whiter, and cooler soot is redder, just as is true for heated metal. At typical campfire and candle temperatures, the burning soot glows with a pleasing, soft yellow color.

One can imagine Davy explaining this unexpected finding to his eager protégé Michael Faraday. Thirty-five years later Faraday explained during one of his candle lectures, "You would hardly think that all those substances which fly about London, in the form of soot and blacks, are the very beauty and life of the flame."

Many people know that soot is made of carbon, but few are aware of the soot's atomic complexity. A soot particle is no mere speck of graphite or fleck of carbon. Instead it is a marvelously tiny globe made of long ribbons of carbon atoms, wound up like a ball of string. These polyhedral structures float upward in the candle flame, brilliantly lucent as they react with oxygen at the flame's border to form carbon monoxide and carbon dioxide. Besides giving off a lot of light, sooty fires give off a lot of smoke. So, a smoky fire is a bright fire as well.

Soot is a by-product of many combustion processes. While often viewed negatively because of its reputation as a source of air pollution, soot has also played an important role in the advancement of human culture. For example, the cave dwellers of Lascaux wouldn't have seen a thing without the light coming from the burning soot in their lamps. The first inks developed millennia ago in China and India consisted largely of soot particles suspended in liquid. Even today, many manufacturing processes use railcar loads of soot and its close relative, carbon black, as raw materials for a number of important products ranging from photocopier toner to truck tires.

Besides heat, the other principal attribute of fire—luminosity—is the product of burning soot. The brightness of a fire (or for that matter, any light-producing thing) is a commonly desired measurement. Two scientists, Count Rumford and Robert Bunsen, came up with ingenious ways to measure brightness: they invented devices called photometers.

Although humans are not naturally adept at measuring the intensity of light in absolute terms, we are pretty good at comparing the brightness of one source with that of another. That fact is the basis for a device called a photometer, which uses a light source of known intensity to determine the brightness of another. Count Rumford used the fact that the intensity of light diminishes in proportion to the square of its distance from the source to build one of the first photometers. He recognized that the intensity of light or other linear waves radiating from a source, such as a candle or small lightbulb, is inversely proportional to the square of the distance from the source. Thus an object (of the same size) twice as far away receives one quarter the energy. An object four times distant receives only one sixteenth.

Rumford knew that light radiates from a point source in all directions in straight lines. As these rays move farther from the starting point, they separate more. Because the rays separate more, a lesser volume of light impinges upon objects farther from the light source. This phenomenon is called the law of inverse squares. **Diagram 13.1** illustrates this.

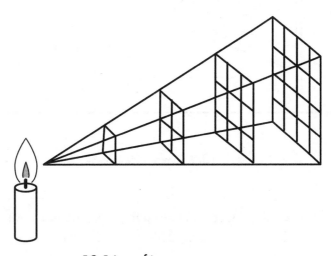

13.1 Law of inverse squares

Rumford Photometer

The Rumford photometer works by moving light sources closer to and farther from a neutrally colored surface until the color and intensity of the shadows cast are the same. **Diagram 13.2** shows the setup of a Rumford photometer.

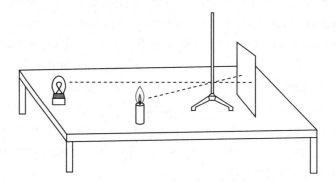

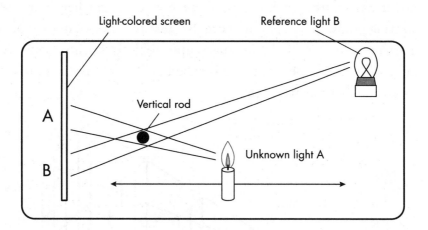

13.2 Rumford Photometer

www.Project11.ThePracticalPyromaniac.com

Materials

- ❯ Scrap wood to make a stand
- ❯ 24-inch dowel
- ❯ Light-colored posterboard
- ❯ 4-watt lightbulb
- ❯ Lamp with shade removed (or other power source for the lightbulb)
- ❯ Candle
- ❯ Match or lighter
- ❯ Tape measure

1. Use the scrap wood to build a simple stand for the wooden dowel so that it stands straight vertically on one end.
2. Prop up the light-colored posterboard at one end of the table and place the dowel about 12 inches in front of it.
3. Set up the 4-watt lightbulb and lamp at one corner of the side of the table opposite the posterboard. Put the candle at the other corner and light it.
4. Darken the room and move the candle, light A, toward or away from the posterboard until the two shadows cast on the screen are of equal brightness.
5. Turn the room lights back on and measure the distances from each light source to the posterboard.
6. Determine the luminosity of the candle. In this example, the known light source is a 4-watt lightbulb whose luminosity is known to be 14 lumens. The distances from the known and unknown light sources to the shadow each casts should measure to be 16 and 13 inches, respectively.

   An equation relating the known distances and reference light-bulb intensity to the unknown light can be set up as follows:

$$\frac{\text{Unknown brightness of candle flame}}{\text{Known brightness of 4-watt bulb}} = \frac{(\text{Distance of candle shadow at equality})_2}{(\text{Distance of bulb shadow at equality})_2}$$

If you remember a bit of your junior high school algebra, it's easy to calculate the candle brightness by plugging in the quantities you know, rearranging the equation, and then cranking it through your calculator to find the answer:

$$\text{Candle brightness} = \frac{(\text{Distance of candle shadow at equality})^2}{(\text{Distance of bulb shadow at equality})^2}$$

$$\times \text{ known brightness of 4-watt bulb}$$

$$\text{Candle brightness} = \frac{13^2}{16^2} \times 14$$

$$\text{Candle brightness} = 9.2 \text{ lumens}$$

We find that a typical candle flame has a brightness of just over 9 lumens. Compare that with the 1,700 lumens put out by a 100-watt lightbulb and you'll soon understand why Edison's invention was so important: you'd need 189 candles to produce the same amount of light as a single 100-watt bulb.

BUNSEN'S PHOTOMETER

Robert Bunsen, inventor of the eponymous laboratory burner, was one of the great scientists of the second half of the 19th century. He leaped to prominence in the chemistry world as one of the few chemists daring to experiment with cacodyl. A poisonous, oily liquid with a pungent and nauseating odor, cacodyl is known for its violent, spontaneous combustion in dry air. It's such nasty stuff that it was briefly considered for use as a battlefield weapon, a sort of poison gas/napalm hybrid. To some chemists, however, its importance and unique properties outweighed its riskiness.

Cacodyl was one of the first organometallic compounds discovered. Organometals are chemical mutants at the intersection of organic and inorganic chemistry. In the 19th century, organometallics were of immense interest because their reactivity made them important in chemical synthesis. For that reason, organometallics attracted ambitious, determined chemists who wanted to make their mark and were willing to assume quite a bit of risk during their investigations.

Robert Bunsen was such a chemist. Born into a family of German academics, he was a professor at the University of Marburg. While best known for his eponymous laboratory burner, Bunsen was so much more than a mere inventor of laboratory equipment. He was the discoverer of the chemical elements cesium and rubidium. He discovered the first, and perhaps still best, antidote for arsenic poisoning, and he somewhat incongruously performed some of the earliest geological studies of Icelandic geysers. A chemist's chemist, Bunsen stood out even among scientific giants for his pioneering work in chemical analysis and chemical spectroscopy and his ability to teach and mentor promising young chemists.

In 1841, he was promoted to full professor at Marburg because of his groundbreaking research on cacodyl. Also called "Cadet's liquid," cacodyl is produced by distilling arsenic and potassium acetate. At the time of Bunsen's research, the nature and chemical composition of the stuff was not fully understood. However, it was well known that cacodyl and its related compounds were poisonous and highly flammable and had an incredibly nauseating odor even in the tiniest of quantities. One of Bunsen's research projects produced a cacodyl-related compound that he named "Alkarsin." He seemed to take some ironic pleasure in describing the Alkarsin's effects in detail, probably because it was such incredibly nasty stuff.

It is a clear, colorless liquid, heavier than and not miscible with water. Its odor is extremely repulsive, reminiscent of the odor of arsine. Even small amounts are extremely lachrymatory [tear-gas like] and cause an almost unendurable, long-lasting irritation of the nasal mucous membranes. Exposure to the vapors for a longer time causes nausea and oppression of the chest. The liquid causes violent itching when in contact with the skin. Its taste is similar to its odor and it is very toxic. On exposure to air or oxygen, thick white fumes are evolved in an exothermic reaction, resulting in a pale flame with formation of water, CO_2, and As_2O_3, the latter as a white smoke.

Few scientists were as courageous as Bunsen in pursuit of knowledge and discovery. These were the days before safety glasses and fume hoods were standard issue, and chemists were the daredevil motorcycle jumpers of the scientific community.

Robert Bunsen would live for days in a dark room, preparing his eyes for the subtle work of comparing very small differences in colors and shading. But such an ordeal made even the workaholic Bunsen seek a better way to the end he sought. Ultimately, he designed a better photometer, one that provided more precise and faster data. It was called the grease spot photometer.

This invention was a game changer in the world of optics—a step up from Rumford's photometer in terms of accuracy and ease of use. An ingenious but simple device, it is easily made by putting a drop of grease on a piece of unglazed white paper and arranging the paper vertically between the two lights to be compared, with one face turned toward each light. The grease spot makes that part of the paper more transparent than the rest and allows more of the light to pass through the paper. If the paper is lit from only one side, the grease spot will seem darker than the rest of the paper when looked at from the lit side and lighter when looked at from the unlit side.

When the intensity on both sides is exactly equal, the grease spot is no longer dark but instead appears to be of the same brightness as the rest of the paper. The grease spot virtually disappears! When that happens, the relative intensities of the two lights can be calculated from their distance from the paper using the same equation as used with the Rumford photometer.

$$\text{Brightness of candle} = \frac{\text{Distance of candle to grease spot at equality}^2}{\text{Distance of reference light source to grease spot at equality}^2}$$

$$\times \text{ known brightness of reference light source}$$

"Had Bunsen been unknown in other branches of science, this simple grease spot would alone have been sufficient to have handed his name down to posterity," wrote one late 19th-century scientist. Other less scientific types were impressed as well. When Emperor Frederick of Germany was shown Bunsen's photometer and the principle of use explained, His Imperial Highness remarked, "For the first time in my life I now know the value of a spot of grease."

The Bunsen Photometer

It's easy to build your own Bunsen photometer to accurately measure the intensity of a flame. All it takes are a few PVC pipes and fittings, a clamp, and a night light.

www.Project11.ThePracticalPyromaniac.com

Materials

- (4) ½-inch PVC tee fittings
- (2) ½-inch PVC elbow fittings
- (2) pieces of ½-inch PVC pipe, 2 inches long
- (1) piece of ½-inch PVC pipe, 3 inches long
- (1) piece ½-inch PVC pipe, 36 inches long
- Glue
- (1) 4-watt, 14-lumen night light (Verify the lumen rating with the manufacturer, if possible.)
- (1) ½-inch-diameter, 6-inch-long taper candle
- Small amount of grease
- (1) 3 × 3-inch piece of stiff white paper
- (1) 3-inch C-clamp
- Transparent tape

1. Arrange the PVC pipe and pipe fittings as shown in **diagram 13.3**. (There is no need to cement the PVC pieces in place permanently.)

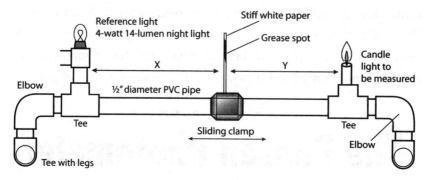

13.3 Bunsen Photometer

2. Glue the night light to the top of the 4-inch piece.
3. Place the candle in the open end of the tee fitting on the end opposite the night light.
4. Place a drop of grease in the middle of the paper and tape the paper to the C-clamp.

TO MEASURE THE BRIGHTNESS OF A CANDLE

1. Light the candle, turn on the night light, and darken the room.
2. Move the C-clamp back and forth on the 36-inch PVC pipe, keeping the grease-spotted paper perpendicular to the pipe. At some point the light intensity from the candle and the light-bulb will be equal and the grease spot will seem to disappear. Tighten the clamp and note the position.
3. Measure the distances from the clamp to the lightbulb and the candle. Plug the distances into the formula, using 14 lumens for the intensity of the lightbulb. (Check with the bulb manu-facturer to confirm its lumen rating.)

$$\text{Brightness of candle} = \frac{\text{Distance of candle to grease spot at equality}^2}{\text{Distance of reference light source to grease spot at equality}^2}$$

$$\times \text{ known brightness of reference light source}$$

14

THERMOCOUPLES

By the end of Michael Faraday's tenure at the Royal Institution, much of the mystery surrounding the nature of fire had been solved. Lifetimes of solid research by some of the best scientists in the world had broken down the 2,000-year-old walls of ignorance. By the middle of the 19th century, scientists understood fire's chemical and physical processes, at least at a macroscopic level. The scientific community then turned its attention from the theoretical aspects of fire to the practical ones.

For example, consider the most commonly measured or controlled parameter, that of temperature. We wait for it on morning weather reports, we dial it in on household thermostats, and we set it on the oven to get the desired results on the food we cook. Everybody has a qualitative, inborn understanding of hot and cold. Hot coffee is better than tepid coffee, and cold beer (to everybody but the British) is better than lukewarm beer. But the ability to precisely measure temperature is an accomplishment of only the last several hundred years of human civilization, and the ability to measure more extreme temperatures, such as those occurring within various zones of Faraday's candle, is even more recent.

Until about 400 years ago temperature measurement was subjective and qualitative. Temperature scales as we know them today did not exist. The medieval temperature taker might be able to tell if the temperature of an object was higher or lower than something else, but attaining much accuracy was difficult, if not impossible. A blacksmith would know how hot an iron bar was by the color of its glow. A candle maker could gauge the temperature of the tallow by its relative softness. For less-defined jobs, only the most general determinations of temperature could be made: is the material in question hot enough to cause water to sizzle, does it melt sulfur, or does it cause beeswax to puddle?

The first significant advances in the ability to quantitatively measure temperature occurred in the late 16th century. Galileo, the famous Italian scientist, filled a tube with fluid and a number of colored glass globes, each with a slightly different density. As the fluid warmed or cooled, its density decreased or increased, causing the sealed glass globes to sink or float. By calibrating the different densities to different temperatures, Galileo quantitatively assessed temperature. However, this brought up a new question: how to correlate those floating objects to a universal temperature scale, a number that all people could understand and relate to.

Remember, there was nothing like the Celsius or Fahrenheit scales at that time. All Galileo could say was that the temperature of the fluid correlated to the floatation of an object of some specific density. He might understand that, but no one else did. To remedy that, Renaissance-era scientists began working on a universal temperature scale.

By the early 1700s, perhaps 30 different scales were in use. In 1714, Daniel Fahrenheit invented both the mercury and the alcohol thermometers. Fahrenheit's mercury thermometer consisted of a thin, sealed tube of mercury with no air in the tube above the mercury pool. That allowed the mercury to expand and contract without having to compress any air in the space above it. In addition, since it was a closed tube, Fahrenheit's thermometer was free from the distorting effects of atmospheric pressure. For those reasons, Fahrenheit's device worked particularly well and gave him so much clout and recognition that the odd temperature scale he devised, with water freezing and boiling at the decidedly nonround numbers of 32 and 212, is still in widespread usage.

Fahrenheit's closed mercury tube thermometer was very useful for measuring within the range of temperatures in which humans work and live, but it is not suitable for measuring the 500-degree Celsius temperatures found within the oxidizing reactions of fire. For those temperatures, a different method was required. That method would not become available until scientists better understood the phenomenon of electricity.

THE THERMOCOUPLE

Prior to 1800, the only way to store and apply electricity was through the use of a Leyden jar. The Leyden jar is a primitive but large capacitor that discharges its entire store of electricity at once, in one big (and sometimes painful) flash of voltage. While electricity was an amusing phenomenon, there could be little practical use for it until someone found a way to control the amount of current over a sustained period. That's why there was great excitement in the scientific community when, in 1800, Alessandro Volta built a device comprised of alternating zinc and copper discs between pieces of cardboard soaked in saltwater. This "voltaic pile," which reliably provided a continuous current, led to the first practical uses of electricity.

Among the first piles large enough to be of practical use was Humphrey Davy's 2,000-cell monster in his laboratory at the Royal Institution. Davy used it to strike the arc of the arc light described in chapter 10.

Roughly 20 years after Volta's innovation, Thomas Seebeck, a German-Estonian scientist, discovered another way to produce electricity. He found that that two different metals joined to form a circle demonstrated increased magnetic properties when he applied heat at the junction between the metals. Michael Faraday, reading about Seebeck's intriguing discovery, began his own investigations into the matter. Faraday connected a bar of antimony to a brass wire and heated the junction point. He wrote that doing so resulted in magnetic effects. "Antimony and brass wire. Bar being heated at one end the north pole of a needle would go round," he jotted down in his notebook. "Effect of the needle (on a compass,) was very decided, powerful, even, and constant." When the great Faraday took notice of this effect, other scientists followed. It was soon determined that besides magnetism, an electrical current was also being induced in the circuit. This came to be called the thermoelectric effect.

In 1885, Henry Le Chatelier took this idea and devised the thermocouple. It was the first commercially practical device capable of measuring a temperature above 500 degrees Celsius. His thermocouples consisted of two wires made from different metals joined together. When the junction of the two metals was heated, an electric voltage was created. The temperature at the junction of the thermocouple was determined by comparing the voltage created to a standard reference table that correlated voltage to temperature.

Constructed of one platinum wire and one platinum-rhodium alloy wire, Le Chatelier's thermocouple provided a usable range and level of accuracy previously unknown. The Le Chatelier device was patented and manufactured in France, allowing for the first time accurate measurements of the high temperatures occurring within oxidation reactions. It was now possible to accurately measure the conditions inside a flame.

Thermocouples are in widespread use today. Most gas water heaters use them to monitor the presence of a pilot light. Engineers and technicians use them frequently because they are inexpensive, simple, and reliable. Like Le Chatelier's invention, the modern thermocouple consists of two different materials joined at one end and separated at the other. The separated ends yield a voltage that is proportional to the heat applied to the junction of the two metal types. The hotter the temperature at the junction, the higher the voltage measured by a voltmeter.

Different combinations of metals are used depending on the temperatures being measured and the conditions in which the thermocouple is placed. Thermocouples made of iron and copper-nickel alloy are most commonly used, but there are dozens of different metals and combinations available for specialized situations. The different compositions are given letter designations (e.g., *K*, *J*, or *S*) and are standardized across the measurement industry.

Thermocouples are simple to use, requiring nothing more than a voltmeter. However, since the voltage signals generated are very small, the meter must be quite accurate, otherwise large errors result. Since the voltages that thermocouples produce are not linearly proportional, voltmeters must refer to tables provided by each thermocouple's manufacturer to determine the temperature.

Measuring the Temperature of a Flame with a Thermocouple

Understanding what a thermocouple is and how to use it enables you to peer inside the workings of nearly any flame. **Diagram 14.1** shows the basic idea of a thermocouple, wonderful in its simplicity but far-reaching in its usefulness to any experimenter who dares explore the nature of fire.

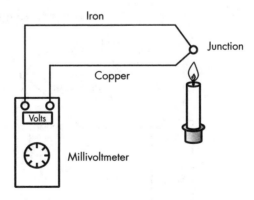

14.1 Basic thermocouple

For this project, we will include a cold junction made of ice water, known to be 32° F, to make the thermocouple reading more accurate.

www.Project11.ThePracticalPyromaniac.com

Materials

- (2) unsheathed (bare wire) fine-gauge Type K thermocouples, 12 inches long (Wire diameter of 0.01 inches is a good size, although other diameters will work.
- (2) copper wires, 12 inches long, with an alligator clip at one end of each
- Voltmeter with millivolt range scale
- Glass or plastic container of ice water
- Type K Thermocouple Reference Chart (These typically come with the thermocouple when purchased, or you can find one online at http://www.temperatures.com/tctables.html)
- Screwdriver
- Candle
- Long-handled matches

1. Splice the two thermocouples back to back so that the aluminum alloy wire of one is attached to the aluminum alloy of the other. You now have one continuous thermocouple assembly with two sensing junctions. Refer to **diagram 14.2**.

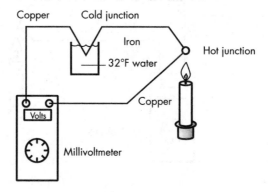

14.2 Error corrected thermocouple

2. Splice the bare end of the copper wires to the nickel alloy ends of the thermocouple.
3. Place the thermocouple junction that is directly connected to the negative voltmeter probe in the ice water container. This is the cold junction.
4. Attach the voltmeter probes to the alligator clips on the copper wires.
5. The thermocouple is ready for use, and the hot junction can measure temperatures as hot as 2,500°F.

USING THE THERMOCOUPLE

Whenever two dissimilar metals are joined as depicted, a voltage exists between the two open ends. If you measure the voltage and look it up in a reference table that correlates temperature to voltage levels for type K thermocouples, voila—you now know the temperature at the junction!*

1. To use the thermocouple, place the hot junction in the area in which the temperature is to be measured. Because the wire is so fine and the junction itself is hard to see, it will take some practice handling the thermocouple to get steady readings on the voltmeter.
2. From the readout of the voltmeter, note the voltage in millivolts.
3. Look up the voltage on the thermocouple reference chart that came with the thermocouple and find the corresponding temperature.
4. The small size of the thermocouple makes it possible to accurately measure the temperature at different points within the flame. By doing so, you can quantify the information found in chapter 2 using the heat map.

---

*It's actually a bit more complicated than that because the introduction of a voltmeter into the circuit creates an error unless compensation is provided. But for our purposes, the error is small enough to ignore.

Build a Thermocouple from Scratch

www.Video7.ThePracticalPyromaniac.com

Commercial thermocouples are cheap, accurate, and easy to use. But hardcore do-it-yourselfers may be interested in constructing a thermocouple from scratch. It is possible to make a useable temperature-measuring device; however, don't expect to obtain the accuracy attainable with manufactured products.

Because we're not terribly concerned with precision, we'll ignore the use of a cold junction and use the homemade thermocouple to measure relative, instead of absolute, temperatures.

www.Project11.ThePracticalPyromaniac.com

Materials

- ◉ (1) 12-inch-long piece of 18-gauge bare aluminum wire (commonly found in hardware stores)
- ◉ (1) 12-inch-long piece of 18-gauge bare copper wire (commonly found in hardware stores)
- ◉ Voltmeter (This thermocouple puts out only 3 millivolts at most, so be certain your meter has a range that includes this level of sensitivity.)
- ◉ Candle
- ◉ Matches or lighter

1. Twist together the ends of the aluminum wire and copper wire. Lo and behold, you've made a thermocouple.
2. Connect the voltmeter probes to the open end of each wire.
3. Light the candle.
4. Place the metal junction in the edge of the flame, at the spot where we found the highest temperatures using the heat map in chapter 2. The voltmeter should display a small but steady voltage reading.
5. Move the metal junction to the center of the candle flame, touching the top of the wick. The voltmeter now displays a lower voltage because the center of the flame is cooler. If you are ambitious, record the voltage and heat combinations on a spreadsheet and use the values to create a reference table for your thermocouple. Note that our simple copper/aluminum couple does not generate much voltage, so it's mainly useful for measuring high temperatures with relatively low precision.

⚠ 15 ⚠

TECHNICOLOR FLAMES

Many chemical reactions, including the combustion process we term fire, produce energy in two forms: heat and light. In this chapter we continue our examination of the way fire produces light that we began in chapter 13.

Light is actually electromagnetic radiation. It results from either black-body radiation or molecular or atomic emissions. In this chapter, we'll look at fire light produced by both processes. Black-body radiation is more commonly termed "incandescence" and refers to the light produced as a consequence of an object's temperature. Within a fairly narrow band, an object's temperature produces radiation that the human eye can perceive. Higher-temperature objects radiate energy at relatively short wavelengths and the visible light generated by them tends toward blue and violet. Lower temperature objects generate longer wavelengths and so appear red and orange.

The maximum temperatures generated in a candle flame are toward the high end (nearly 2,600°F), so the flame is blue at the bottom edge where the combustion is most complete, the temperature is hottest, and no soot obscures the pure flame.

In the part of the candle flame above the blue light, combustion is not so complete. Within the flame envelope, tiny particles form. They are carbon based soot and molecules of carbon and hydrogen compounds including carbon monoxide, acetylene, and ethyne, to name a few. They break loose from the wick, fly around, heat up, and incandesce, but at a lower temperature than the hydrocarbons that produced the blue light. That's why the flame is orange-yellow.

In earlier chapters, we described a candle flame as the result of several complex chemical reactions. The reactions have several parts, beginning with the liquefaction and transport of paraffin hydrocarbons from the wax bowl to the wick's tip, where the wax is vaporized. From there, the gasified fuel heats up and moves outward to the edge of the flame. There, it combines with oxygen from the surrounding air and reacts, producing the most intense heat.

A lit candle, which physicists would call a diffusion flame, may look simple but is actually mind-bogglingly complex. The movement of gases and fuels inside the moving flame, while certainly not random, is incredibly hard to model, as it's caused by myriad factors. These include (1) the breakdown of long wax molecules in a host of smaller fuels, (2) the diffusion of those particles inside the moving flame, (3) the always-changing relative concentrations of fuel and oxygen, and (4) the intricate mechanics of the capillary action by which candle wax wicks and then vaporizes to become burnable fuel. So while we appreciate its bright, warm glow, understanding a candle's flame past a superficial level is challenging.

However, there is a flame that's a bit (but only a bit) easier to understand: the one invented by Robert Bunsen, the Bunsen burner. Instead of the complex, nearly chaotic movement of air and fuel molecules from center to edge and from bottom to top, a Bunsen burner flame is neat and orderly. The fuel molecules in propane and methane gases are much smaller and simpler than the big and complex molecules in paraffin. In a Bunsen burner, adjustable air-entraining holes in the burner's metal torch premix the fuel in precise proportions. This creates a flame with a steady, even shape.

The color of the premixed flame is a nearly transparent blue, which signifies that few unburned carbon particles and little soot is present. If the air and fuel supply remains constant, so does the flame; flickering is not an attribute of a well-tuned Bunsen burner flame. Because of the simplicity, efficiency, and adjustability of the Bunsen burner flame, it is a chemist's best friend.

In 1843, Bunsen was hard at work in his laboratory examining the chemical structure and other important characteristics of cacodyl and its related compounds. Finding the odor intolerable, Bunsen invented an elaborate, unwieldy, but workable face mask. The mask—basically, a glass face shield attached to a long, flexible tube with an open end snaking out the window for fresh air—was adequate protection from the poisonous fumes.

However, cacodyl auto-ignites upon exposure to air. While he was working in his laboratory one day, a sample of the stuff exploded violently, shattering his mask, hurling glass shards in his face, and permanently blinding his right eye. Down but not out, he continued his chemical investigations. He soon had another, even closer brush with death. After he inhaled the vapors of cacodyl chloride, he hovered on the brink of death for several days. When he finally recovered, he decided he had enough and moved on to tamer investigations. It is lucky for those interested in the physics of fire that he did. Bunsen's subsequent work in chemical analysis and spectroscopy led directly to understanding the relationship between fire and color.

Before the Bunsen burner, the lamps and burners used in laboratories produced soot-filled, richly colored, and hard-to-control flames. Such dirty flames made it impossible to study the colors produced when elements and their compounds were heated. From the work of Lavoisier, Priestley, and others, Bunsen knew that oxygen was essential to combustion and that incomplete combustion caused soot and obscured flame colors. Bunsen determined that the way to obtain clear, clean fire in the laboratory was to mix the fuel gas with air in just the right proportions before igniting it.

Bunsen and his assistant, Peter Desaga, built a burner consisting of a hollow metal cylinder with holes bored in strategic locations. Through these adjustable holes, air could enter and premix with the fuel gas, providing the investigation-friendly flame that chemists so keenly desired.

Bunsen opened the door to the most basic and important method for determining the components of a chemical mixture: flame analysis, often called the flame test. Unlike the orange flame of a candle or the yellow clouds of an alcohol lamp, the pale flame from a Bunsen burner is nearly perfect for examining the unique colors different elements produce when heated.

The flame test identifies the components of a substance or mixture. In its simplest form, the substance to be tested is placed on a thin wire then moved into a Bunsen's burner flame. The colors produced by the flame test are compared to known standards to confirm the presence of individual elements in the sample.

Bunsen flame-tested sodium chloride (ordinary salt) and noted the resulting bright orange-yellow glow. He also observed the same color in the flame of sodium bromide, as well as any sodium compound placed in the flame. Bunsen discovered rubidium and cesium by noting the characteristic colors those elements produced in his flame test. Other scientists used Bunsen's burner and flame test to make significant discoveries. Among them was William Crookes, the young scientist who transcribed Faraday's Christmas lectures.

Born in London in 1832, Crookes was the oldest of 16 children of a wealthy tailor and real estate investor. At age 16 he entered the Royal College of Chemistry in the hopes of studying organic chemistry. While there, he took a job as assistant to one of Germany's most distinguished chemists, August Wilhelm von Hofmann. This role was pivotal, as von Hofmann's prestige allowed Crookes to attend meetings at the Royal Institution.

Crookes, who could afford his own private laboratory located not far from the Royal Institution, decided to apply Bunsen's flame test to various ores and compounds after reading an article on Bunsen's work in an academic chemistry magazine. Crookes wrote, "With so delicate a reaction, the presence of elements existing in so small quantities as to entirely escape ordinary analysis, may be rendered visible."

Crookes identified arsenic's close cousin, the element thallium, when he heated thallium-containing ore in the flame of a Bunsen burner and found it produced a unique spectra (or light signature), in this case a distinctive bright green line. The Greek word *thallos* means "green twig."

- - - - - - - - - -

The Flame Test

www.Video8.ThePracticalPyromaniac.com

The flame test is used to identify an unknown metal compound by comparing the color the compound turns in the flame of a Bunsen burner with the colors in a chart. The intense heat of the burner flame ionizes and excites the metal atoms, causing them to emit light at specific wavelengths that the human eye interprets as colors.

www.Project12.ThePracticalPyromaniac.com

Materials

- ❯ Wire snips or cutters
- ❯ Coil of thin (26-gauge or even thinner) nichrome wire
- ❯ (2) needle-nose pliers
- ❯ Container of distilled water
- ❯ Bunsen burner or propane torch
- ❯ Chemicals to be tested (Chemicals that work well include table salt, strontium nitrate, copper chloride, lithium sulfate, boric acid, calcium chloride, and potassium chloride. See chapter 1 for sourcing materials.)

1. Using the cutters or wire snips, cut a piece of nichrome wire about 5 inches long. Using the pliers, bend one end of the nichrome wire into a small loop about $3/16$ inch in diameter. Avoid touching the metal loop because the sodium from the salt on your skin will interfere with other, less reactive elements.

2. Dip the loop into a container of distilled water. Shake off the water.

3. Light the Bunsen burner or propane torch and place the metal loop in the flame. If the flame does not change color, the loop is clean and may be used. If the flame changes to a different color, the loop is dirty and must be cleaned before continuing.

4. Once the loop is clean, dip it in distilled water again.

5. Dip the clean loop into the chemical powder to be tested. You only need a small sample to adhere to the loop.

6. Insert the loop into the clearest part of the Bunsen burner or propane torch flame. Note the flame color, which indicates the presence of elements according to the list below.

COLOR CHART

Red	Lithium or Strontium
Yellowish Red	Calcium
Yellow	Iron or Sodium
White	Magnesium
Greenish White	Zinc
Green	Copper or Boron
Blue	Lead, Selenium, Bismuth, or Cesium
Purple	Potassium or Rubidium

7. Let the nichrome loop cool and then discard it. (You can try to reuse the wire loop, but you'll find it nearly impossible to clean completely and will end up contaminating the next sample you try to analyze.)

The flame test is fun and easy and provides a good example of how chemists use fire to analyze unknown compounds. However, there are limitations to this method. Sodium, even in low concentrations, burns so brightly that even a little bit masks other elements. If you see the bright yellow flame of sodium, be aware that other elements may be present but hidden in the sample. Some metals produce almost identical flame colors. For instance, the beautiful red flames of lithium and strontium are very difficult to differentiate. Low concentrations of metals (except for sodium that glows bright yellow even in low concentrations) may be insufficient to color the flame.

- - - - - - - - - -

The Stoplight of Fire

www.Project12.ThePracticalPyromaniac.com

Materials

- ◉ Rubber gloves
- ◉ Safety glasses
- ◉ 2 grams strontium chloride (See chapter 1 for suppliers.)
- ◉ (3) 6-ounce Pyrex or other heat-resistant custard cups
- ◉ 1 teaspoon distilled water
- ◉ Glass stirring rod
- ◉ (3) strips of thin, white 100 percent cotton cloth, each 3 × 6 inches
- ◉ 2 grams sodium chloride (table salt)
- ◉ 2 grams copper chloride
- ◉ Methanol (such as Heet brand gas line dryer)
- ◉ Long-handled lighter

1. Put on rubber gloves and safety glasses.
2. Place the strontium chloride in a clean custard cup and add 1 teaspoon of water.
3. Stir with a clean glass stirring rod until the chemical dissolves.
4. Soak up the solution using the cloth. Make sure the solution is evenly distributed throughout the cloth.
5. Unfold the cloth, set it on a clean, stain-resistant surface, and let it dry completely.

6. Repeat the procedure for the sodium chloride and the copper chloride.

7. Clean the custard cups thoroughly and dry them.

8. Soak each chemical-infused cloth in 1 teaspoon of methanol and place each strip in its own custard cup, arranged so that an edge of the cloth is exposed for easy lighting.

9. Arrange the custard cups in a row with the sodium chloride-treated cloth in the middle.

10. Dim the lights and ignite each cloth in rapid succession with the long-handled lighter. The strontium-treated cloth burns with a deep red flame, the sodium chloride with a yellow flame, and the copper chloride with a vivid green-blue flame, like a flaming stoplight!

⚠ 16 ⚠

THE FIRE TORNADO

English, being a flexible and exceedingly nuanced language, has a huge number of words and phrases to describe the high-temperature, self-sustaining chemical reaction that is the subject of this book. If it's a small fire, a single syllable will do; it's merely a glow, flame, or a spark. A bigger fire requires an additional syllable, for now things are burning, ablaze, or perhaps even flaring. Describing a huge fire requires even more syllables; we now speak of a conflagration, an inferno, or even a holocaust. Such darkly connotative words scare up images of uncontrollable destruction and devastation.

However long or descriptive, words cannot describe the rivers of flame that roared through the dry, dusty center of the United States on October 8, 1871. On that single day, a pack of the most destructive and deadly fires in the history of the United States rampaged through Illinois, Michigan, and Wisconsin, turning a significant portion of those states into an Old Testament version of hell on earth. The fires turned a million and a half acres of Midwestern verdancy into pyrolized blackness and killed thousands of people. The devastation was so extensive that the exact number of victims is not known. Many bodies were incinerated, and the local government records of inhabitants were burned to ashes. What is known with certainty, though, is that the death toll was staggering.

Early on the morning of October 8, 1871, a meteorological depression stalled over western Kansas. Extending from Nevada to Ontario, a massive weather system produced an unusually widespread area of low pressure that kicked out sustained high-velocity winds along its broad eastern edge. The previous summer and autumn months had been drier than normal, leaving many areas of the Midwest extremely parched. The drought, combined with the slash-and-burn land-clearing practices of the time, provided perfect conditions for the establishment of numerous small brush and forest fires.

When the winds from the central prairie's weather system kicked in, a perfect firestorm erupted, engulfing an area larger than the size of Delaware in a terrible conflagration. Those who lived through it never forgot the night and day that followed.

— — — — — — — — — —

Chicago in the 1870s was one big fire hazard. Its buildings, its bridges, even its sidewalks were constructed almost entirely of wood. The science of fire prevention had not yet been developed, so most people lived and worked in places that by modern standards would be considered ramshackle tenements and deadly firetraps.

The fire, according to the well-known but probably untrue story, started when a cow kicked a lantern in Catherine O'Leary's barn on De Koven Street on the west side of Chicago. Within minutes, the flames, fanned by winds generated hundreds of miles to the west in the Kansas cyclone, roared south and east toward downtown Chicago. Homes, factories, and warehouses quickly caught fire, sending flaming debris eastward on the roaring winds.

Observers in Chicago noted that the tallest parts of buildings caught fire first as a rotating, upward current of flame carried burning faggots and cinders long distances. Trees went up like matchsticks. Horses kicked down stable doors to escape, and those that could not bellowed in agony. People fled without taking time to collect any possessions, running one step ahead of the smoke and flames. If they were lucky, only their backs and hair were burned. Frantic residents ran into Lake Michigan, where they hoped the water would protect them from the rapidly approaching flame front.

For two days the fire roared out of control. The city had a professional fire department of only 185 men, and those men were tired, having fought fires throughout the city on a nearly continual basis for the two weeks preceding the big blaze. They fought a good fight, but when it was over, four square miles of the most built-up portion of the city had been devoured by the flames. Nearly 300 people were dead, and 100,000 residents had lost their homes. Seventy thousand buildings were destroyed.

Yet as bad as the Great Chicago Fire was, far worse fires burned elsewhere on the same day. A few hundred miles north of Chicago, in the mixed conifer and hardwood forests, another, much larger firestorm developed. Here, the epicenter was not densely packed, urban Chicago, but rather a small, prosperous logging and railroad town on the shores of Lake Michigan: Peshtigo, Wisconsin.

The summer and autumn in northeastern Wisconsin had been as dry as elsewhere throughout the Midwest. Practically no rain had fallen between the beginning of July and the beginning of October. Rivers had dried up, swampland crackled underfoot, and the resinous tops of the area's great pine forests were like match heads waiting for a light.

Around dinnertime on Sunday night Peshtigo residents heard rumbling off to the southwest, a sound that survivors said was similar to the movement of a great number of freight trains. People were puzzled by what it could be, but not for long.

Within an hour, a great orange flare appeared in the western sky, and all manner of things on fire began dropping from above. The town's buildings, constructed in the same manner as those in Chicago, burst into high licks of flame. Town residents close enough to get to the Peshtigo River before the onrush of the fire dove into the water in a frenzied bid for safety. Immersed in the slow-running current, people still fought for their lives:

Men, women, and children filled the stream, the women holding babies under water to their necks, and wetting their heads to prevent burning—for the heat was like that of a blast furnace. Many had the hair burned from their heads as they lay immersed in water; many others, in the desperate struggle for air, breathed flames and died right there in the watery element. Another danger was the timbers from a burning mill which floated by aflame and had to be fought off by the men. But the most serious threat to the refugees in the water was from the livestock, for cattle, horses, swine, all frantic with fear, rushed into the stream trampling the helpless women and children, many of whom were thereby lost who might otherwise have been saved.

The fire consumed an area of pine and oak forest 30 miles long and about 15 miles wide. Gone in less than two hours were forest, homes, sawmills, farms, and people burned to ashes in a firestorm of unbelievable proportions. It seemed to survivors as if "hell yawned and the fire spit out."

FIRE TORNADOES

Fire tornadoes, also known as fire whirls, were a key factor in the extreme deadliness and destructiveness of the October 1871 fires. They form only rarely, when a "perfect storm" of wind, low humidity, abundance of fuel, and bad luck come together. But such were the conditions in Chicago and Peshtigo.

In Chicago, fire crews had a chance at containing the fire early on, but the moment the fire whirls kicked in—carrying burning planks aloft and depositing them three-eighths of a mile ahead of the firefighters—the city was lost. In Peshtigo, the fire tornados were even larger and more destructive. Forming quickly from great heaps of dry organic material in the dried-up swamps on the west side of the area, the fire whirls swooped into the town.

According to the U.S. Forest Service, fire whirls are frequently observed in wild land fires. They are most often small, but large ones may form, depending on atmospheric and fuel conditions. Scientists estimate that the biggest whirls have rotating air speeds that exceed an F-5 tornado, reaching a whopping 300 miles per hour.

Meteorologists would term such a phenomenon a vortex, which is a volume of fluid possessing a rotational motion. Examples of natural vortexes abound: smoke rings from a cigarette, water going down a drain, and dust devils. In fact, a fire whirl and a dust devil are similar in many ways. Both form when a layer of hot air is covered with a layer of cooler air. The hot air, lighter and more buoyant, pokes a hole in the cool air above it and rises up through the opening.

If the conditions are right, the rising hot air begins to spin, and a whirlwind forms. The rotational trigger for the vortex could be as simple as a gust of wind rushing past a rock or small mound of dirt. The difference between a dust devil and a fire whirl is to some extent merely a matter of degree. Where the surface temperature of the dry, dusty earth that typically creates the dust devil can be as hot as 150°F, the fire whirl is created by air temperatures that may exceed 2,000°F. These extreme temperatures create massive columns of rising hot gas that shoot up erratically through the cool air layer. The hotter the fire and the more unstable the atmospheric conditions, the more erratically the hot gas rises. Some small triggering event makes the fire begin to spin as it ascends, the hot gas slipping by and spinning around with the cool air containing it, due to friction. A fire whirl is born.

In the center core of a whirl, the air is drawn down, while on the edge of the vortex there is a strong updraft. The air and hot gas rotate like a hurricane, with the greatest wind speeds closest to the core or eye of the whirl. The movement of the air works like a giant blower or bellows, providing a surfeit of oxygen, which causes the fire to burn with an intensity five to six times greater than a non-rotating fire. Little wonder, then, at the speed with which immense destruction was visited upon the unfortunate town of Peshtigo.

- - - - - - - - - -

Building the Fire Tornado

www.Video9.ThePracticalPyromaniac.com

The Fire Tornado dramatically illustrates the effect air has on a fire's shape, burning rate, and fuel consumption.

KEEPING SAFETY IN MIND

1. This project is for adults or for those under the close supervision of an adult.
2. Keep careful watch on the fire at all times.
3. Remove all combustible and flammable objects from the area. Keep the project away from flammable walls, surfaces, curtains, etc.
4. Do not allow the craft sticks or anything else to ignite.
5. This project creates smoke and should be performed in a very well ventilated location.
6. Keep your fire extinguisher close by.
7. Be sure to read the general safety procedures and disclaimers in chapter 1 before attempting this project.

www.Project13.ThePracticalPyromaniac.com

Materials

- All-purpose fire extinguisher
- Modeling clay, approximately ½ pound
- Turntable (Old phonograph turntables are frequently available at thrift stores or garage sales. Similar devices, such as a lazy Susan, may work as well. The main requirements are that the

turntable spins freely and has a raised edge on which support stakes can be glued.)

- ⊚ Teacup-sized fireproof bowl or crucible
- ⊚ Fast-drying epoxy, or hot glue gun and glue
- ⊚ (12) large-sized craft sticks (approximately 6 inches long × ¾ inch wide)
- ⊚ Piece of aluminum (not plastic or fiberglass) window screen, 36 × 36 inches
- ⊚ 4 straight pins
- ⊚ Tin snips or metal shears
- ⊚ 2 teaspoons kerosene (Do *not* substitute any other fuel, such as gasoline or alcohol.)
- ⊚ Cotton rag, approximately 4 × 4 inches
- ⊚ Long-handled lighter
- ⊚ Heat-resistant gloves
- ⊚ Large fireproof bowl big enough to fit completely over the small teacup-sized bowl or crucible

1. In a well ventilated space with your fire extinguisher handy, use the modeling clay to create a base for the small fireproof bowl or crucible in the center of the turntable. Place the crucible on the clay and press down, so that the crucible stays put when the turntable is spun.

2. Using epoxy or a hot glue gun, glue the craft sticks to the rim of the turntable as shown in **diagram 16.1**. Let dry.

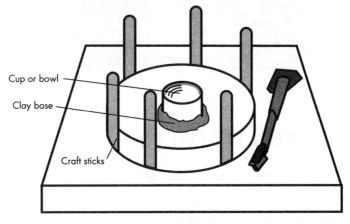

16.1 Fire Tornado base

3. Taking note of the diameter of the circle created by the craft sticks, roll the aluminum window screen into a 36-inch-high cylinder of the same diameter. Push the straight pins through the fabric of the screen along its length to hold the screen in the proper shape. After the screen is pinned into a cylinder shape, use tin snips or metal shears to cut away the excess screen beyond the pins, leaving an overlap of 1 inch.

4. Cut a small diameter hole or flap in the screen just large enough for the long-handled lighter to fit through and extend into the crucible as shown in **diagram 16.2.**

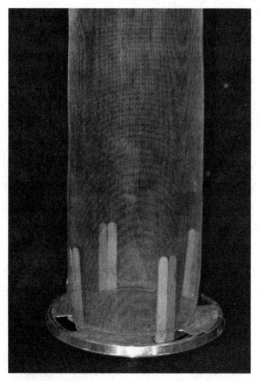

16.2 Fire Tornado assembly

DEMONSTRATING THE FIRE TORNADO

Now that the fabrication of your Fire Tornado is complete, it's time to try it out.

Note: The size and shape of the vortex is best viewed at night.

1. Place 2 teaspoons of kerosene on the cotton rag.
2. Place the rag in the crucible on the turntable.
3. Place the screen cylinder over the craft sticks so that the screen is secured to the turntable when the turntable is spun.
4. Through the ignition hole you cut in the bottom of the screen, ignite the rag with the long-handled lighter. Note the size and shape of the flame in the crucible.
5. Give the turntable a spin by hand or switch the turntable on and set it to the highest available speed. (The faster the rotation, the higher the tornado.) Immediately, the column of flame inside the wire cylinder dramatically grows in height. The fire assumes a spiral shape, similar to a tornado!

16.3 Fire Tornado in action

TO EXTINGUISH THE FLAME

1. Stop the rotation of the turntable. Put on your gloves and remove the screen.
2. Invert the large bowl and place it over the flame, firmly on the turntable surface. This deprives the flame of air and extinguishes it.

FIRE TORNADO PHYSICS

This project clearly shows the effect of air on the size and quality of a flame. A flame is composed of hot gases that emanate upward from the center of the burning materials. As these hot gases travel upward, over the flame, they displace the cooler air above it, which sinks down into the cylinder. Because the cylinder is spinning, centrifugal motion pushes the cooler air toward the rotating screen, making the relative air pressure at the screen higher. But if the pressure is higher at the screen, then it's got to be lower somewhere else, namely in the hot area above the flame. The low pressure there allows the hot, ignited gases from the burning rag to reach upward more easily, creating the flame vortex of the fire tornado.

🔥 ⚠ **17** ⚠ 🔥

GREAT BALLS OF FIRE

17.1 Flamethrower in action

What book filled with fire-related projects could be considered complete without step-by-step directions for building a personal flamethrower? Certainly not this one. This project, first described in my earlier book, *Absinthe & Flamethrowers: Ruminations on the Art of Living Dangerously,* quickly became one of the most popular ones. To be exact, this project is more correctly considered a propane-based flame cannon rather than a flamethrower, but that's probably quibbling. It creates great balls of fire no matter what it's called.

Before you start constructing your flamethrower, decide if you really want to do this. The risks (serious burns, setting the garage on fire, or a visit from an angry fireman) may not be worth the return. But if you know this and still want to go ahead, then read on.

Start by reviewing chapter 1 on safety. I've made several flame-throwers with no major problems, but remember that things can go wrong even through no fault of your own. If you do attempt this project, you and you alone are responsible for what happens.

- - - - - - - - - -

Build a Propane Flamethrower

www.Video10.ThePracticalPyromaniac.com

KEEPING SAFETY IN MIND

1. Do not operate the flamethrower near combustible materials. Keep animals and children away.
2. Inspect equipment for damage and wear prior to each use.
3. Use only nonmodified, government-approved propane cylinders.
4. This device is for adults only.
5. Keep the propane cylinder level and upright. Don't invert cylinders or lay them on their sides. Cylinder valves must be protected. Never lift a propane cylinder by the valve.
6. Don't use a flame to heat up a gas container to increase pressure.
7. Shut everything down if you smell gas. Immediately shut off all valves. Never use a flame to test for leaks. Instead, use soapy water and look for bubbles.

8. Propane is heavier than air, so it will accumulate in the nozzle and other bowl-shaped or low areas. Be certain your area is well ventilated.

9. Keep all sources of ignition away from cylinders, regulators, and hose.

10. Wear protective gear, including safety glasses and heat-resistant gloves.

11. Have your fire extinguisher close at hand.

12. Be certain to comply with all safety guidelines and local ordinances regarding the use of an open flame.

13. Use only in areas suitable for a device of this kind. The area should be secured to keep children, animals, and adults from entering the area.

14. Use extreme caution at all times. You are using an intense open flame, and disregarding safety practices can have severe consequences.

Now, if you're OK with all that, let's make a flamethrower!

www.Project14.ThePracticalPyromaniac.com

Materials

Flamethrower Assembly

- Gas-rated pipe thread sealing compound
- (1) 2-inch-diameter black iron (BI) pipe, NPT threaded both ends, 24 inches long
- (1) $3/8$-inch-diameter flare fitting to $1/2$-inch-diameter NPT fitting, male both ends
- Miscellaneous reducing fittings to reduce from the 2-inch threaded pipe to a $1/2$-inch-diameter coupling at the nozzle and the $3/8$-inch flare fitting that connects to the propane regulator hose

- ◉ (1) ½-inch-diameter steam whistle valve
- ◉ (1) 20-pound standard government approved propane cylinder, filled
- ◉ (1) Variable-setting, high-pressure propane regulator with a 10-foot hose

Stand Assembly

- ◉ (2) 2-inch-diameter BI pipe nipples, 4 inches long
- ◉ (1) 2-inch-diameter BI pipe tee fitting
- ◉ (1) 2-inch-diameter BI floor flange
- ◉ (4) ¼-20 bolts, 2 inches long

Nozzle Assembly

- ◉ (1) ½-inch-diameter pipe nipple (nozzle holder), 5 inches long
- ◉ (1) ½-inch-diameter to ¾-inch-diameter coupling (the nozzle)
- ◉ Igniter Assembly
- ◉ Propane hand torch
- ◉ Large diameter hose clamps

Tools

- ◉ Pipe wrench
- ◉ Electric drill
- ◉ #7 drill bit
- ◉ ¼-20 tap
- ◉ Miscellaneous workshop tools, including pliers, hammer, screwdrivers, and wire clippers

1. Begin by building the flamethrower assembly. Using gas-rated thread compound on all joints, assemble the pipe, pipe fittings, and the whistle valve to form the flamethrower assembly as shown in **diagram 17.2.**

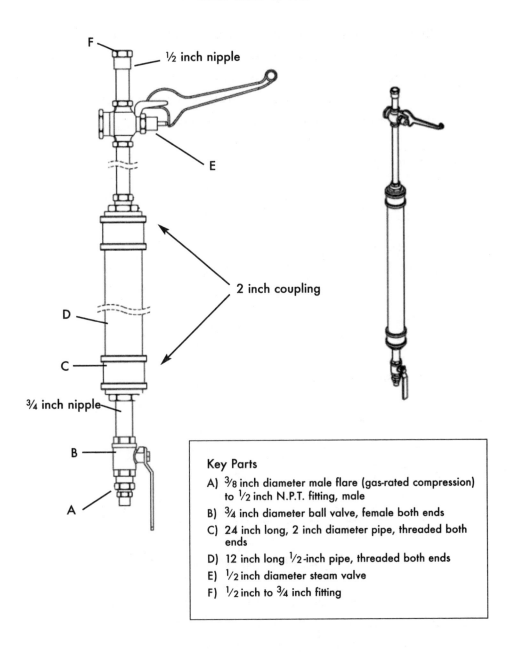

F — ½ inch nipple

E

2 inch coupling

D

C

¾ inch nipple

B

A

Key Parts

A) ⅜ inch diameter male flare (gas-rated compression) to ½ inch N.P.T. fitting, male

B) ¾ inch diameter ball valve, female both ends

C) 24 inch long, 2 inch diameter pipe, threaded both ends

D) 12 inch long ½-inch pipe, threaded both ends

E) ½ inch diameter steam valve

F) ½ inch to ¾ inch fitting

17.2 Flamethrower assembly

2. Build the flamethrower stand. See **diagram 17.3** for details. First, drill and tap holes to accommodate the four 2-inch-long ¼-diameter bolts in one of the 2-inch-diameter nipples as shown in the diagram. Use a #7 drill and a ¼-20 tap to make the threaded bolt holes in the pipe nipples.

All fittings are 2 inches diameter

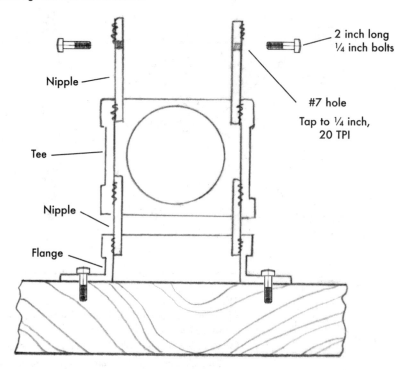

17.3 Flamethrower stand

Refer to **diagram 17.4** for the next three steps.

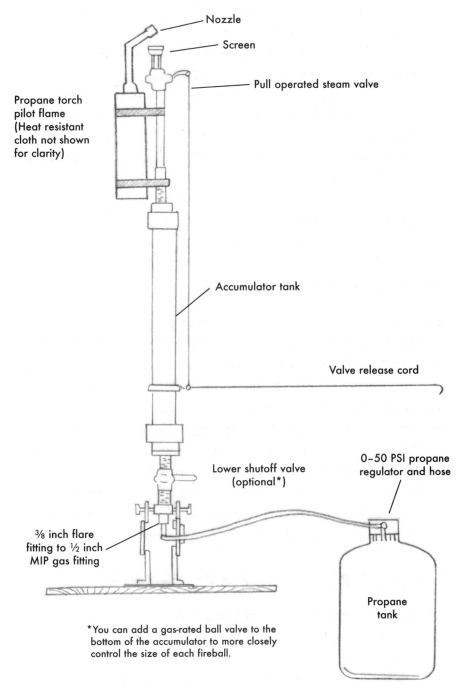

Nozzle

Screen

Pull operated steam valve

Propane torch
pilot flame
(Heat resistant
cloth not shown
for clarity)

Accumulator tank

Valve release cord

0–50 PSI propane
regulator and hose

Lower shutoff valve
(optional*)

⅜ inch flare
fitting to ½ inch
MIP gas fitting

Propane
tank

*You can add a gas-rated ball valve to the
bottom of the accumulator to more closely
control the size of each fireball.

17.4 Fully assembled Flamethrower

3. Next, place the flamethrower into the stand as shown. Position the flamethrower assembly into the flamethrower stand and secure by turning the bolts.

4. Attach the propane pilot torch to the flamethrower assembly using two hose clamps. Position the propane torch nozzle so that the pilot flame extends directly over the flamethrower nozzle.

5. Connect the high-pressure regulator to the propane tank. Connect the ⅜-inch-diameter male flare fitting on the propane hose to the female fitting on the end of the flamethrower handheld assembly.

6. Test the assembly for leaks by checking all connections with soapy water. Bubbles indicate leaks. Repair any leak prior to using flamethrower.

USING THE FLAMETHROWER

Open the propane valve. Ignite the pilot propane torch. Pull the whistle valve cord firmly but slowly enough so the flamethrower does not tip. A large fireball will issue from the nozzle. A small flame, lingering from residual propane in the nozzle, will be present for several seconds. This small flame can be used to ignite the propane for subsequent great balls of fire. To stop, release the whistle valve.

Remember: propane is heavier than air, so non-ignited propane can drift downward from the nozzle and burst into flames suddenly. Therefore, the area underneath the nozzle is a danger zone. Do not stand or have any part of you underneath the nozzle when you operate the device. Stand well off to the side.

SHUTTING DOWN THE FLAMETHROWER

1. Turn off the valve on the propane tank.
2. Lay the flamethrower on its side, so the nozzle is pointing slightly downward.
3. Open all valves except the valve on the propane tank.
4. Use the igniter to burn off the residual propane in the flamethrower assembly. When all propane has left the flamethrower assembly, shut off all valves.

Epilogue

Hollywood filmmakers have long had a penchant for giant, evil-looking insects. The studios started cranking out batches of low budget, campy, but marvelously creative movies in the 1950s and 1960s starring colossal insects. They continue to do so, after a fashion, to the present day. The insects range from the hairy, googly-eyed giant ants of *Them!* to the sometimes good, sometimes bad moth *Mothra* to the insanely violent insectoids of *Starship Troopers*.

However, it's best not to depend on B-movies for your understanding of science and history. Insects, with or without the help of genetic mutations or nuclear radiation, could simply never attain such size. Because of the mechanical properties and limited strength of the crunchy shell, called chitin, that composes the insect exoskeleton, a dragonfly or bumblebee of, say, bald eagle size would plunge headlong into the ground.

In addition, such giant insects simply could not process enough oxygen to live. Insects, insectoids, and spiderlike creatures can grow only so big because they are limited by the degree to which their physiology allows oxygen to reach their vital organs. Entomologists tell us that few land-dwelling or flying insects can grow longer than about five or six inches, because the insects don't have the respiratory system to move oxygen to where it needs to be if the creatures get any bigger.

Millions of years ago, however, the oxygen concentration in earth's atmosphere was unimaginably high by present-day standards, so bugs could get bigger. Much, much bigger. The fossil record is heavy with the petrified remains of huge insects, indicating that levels of oxygen in the atmosphere have changed considerably throughout the earth's existence. Over time this change has had profound effects on the nature of fire.

Imagine you have a time machine that could dial back 300 million years to the time just before dinosaurs. In this period, the Carboniferous Era, the earth is a far different place. The dominant life forms are insects and arachnids. Because of the high oxygen concentration, these guys are huge. As they fly, the nearly pelican-sized dragonflies and mayflies sound like gasoline-powered model airplanes. On the ground, spiders the size of dinner plates run wild next to millipedes three yards long. The biggest scorpions weigh as much as a golden retriever.

With the concentration of atmospheric oxygen so high, fires burn far more energetically than they do today. Infernos kindled by lightning and volcanoes rage unimpeded for months, making even the largest, Peshtigo-sized fires of the modern age seem puny.

Dial the time machine back another 100 million years or so, and things are far different. The earth is devoid of fire. Most people are surprised to learn that fire is a relatively new phenomenon, geologically speaking. In fact, fire from combustion has been a part of nature only since the middle of the Paleozoic Era, or roughly 400 million years ago. Given that estimates of the earth's age generally run toward four billion years, fire has only been present for a small, recent fraction of its total age. For 90 percent of Earth's existence, there was no fire.

Presently, oxygen levels have stabilized, and it makes up around 21 percent of our atmosphere. Fire is an integral part of life on earth, and man-made, not natural, fire constitutes the overwhelming percentage of all burning. Humans have become very, very good at kindling, extinguishing, controlling, and exploiting fire. There are far more fires—in our homes, cars, and businesses—burning away now than at any other time in the history of the planet. But the products of all that fire, the millions upon millions of tons of carbon dioxide produced by our now familiar "high-temperature, self-sustaining, chemical oxidation reaction of hydrocarbon fuels," presents a great problem for our planet that must be solved.

I am an optimist, because I know how genuinely smart and altruistic people of science can be. The turn of the 19th century was a magnificent time to be a scientist. The names and reputations of many of the great explorers of chemistry in general and fire in particular—Faraday, Davy, Lavoisier, Franklin, and others—were well known then and continue to be so today. As John Dalton's funeral in Manchester in 1844 proved, scientists, not entertainers or sports figures, were the heroes of their era. But if the 19th century seemed like a good time to be a scientist, today is far better.

Science, and particularly technology, have evolved into an organized, structured, and cooperative endeavor. It is now the purview of teams of bright lights in research departments and professionally run laboratories. Because of that, scientific progress moves at a velocity magnitudes greater than ever before.

More people than ever want to be or are scientists. Scientists have traded the slim chance of an amateur's fame for the professional's likelihood of a good salary and steady employment. That's a deal most people would take in a heartbeat. The era of the lone genius working in a private lab late into the evening in hopes of discovering a new element has been gone for nearly a century. That is something I lament, at least a bit. For what scientists will the people of the 22nd century talk about when they look back at the close of the 20th century? What names will they still know? Consider the incredible life stories of Lavoisier, Priestly, Rumford, Franklin, Bunsen, and more. One just couldn't make that stuff up.

It's unlikely we will know the name of the individual who solves air pollution, climate change, or global warming, not because there won't be a solution, but because the solution will be the result of thousands of scientists working together. However, these scientists and engineers will all follow in the footsteps of those individual geniuses whose names we do know—the original turn-of-the-19th-century Practical Pyromaniacs who unlocked the secrets of fire.

Bibliography

Brown, G. I. *Scientist, Soldier, Statesman, Spy—Count Rumford.* Boston: MIT Press, 1999.

Davy, John. *Memoirs of the Life of Sir Humphry Davy.* London: Longman, 1836.

Faraday, Michael. *Chemical History of a Candle.* New York: Harper & Brothers, 1860.

Hunt, L. B. "The Early History of the Thermocouple." *Platinum Metals Review* 8, no. 1 (1964): 23–28.

Jackson, Joe. *A World on Fire—A Heretic, and Aristocrat, and the Race to Discover Oxygen.* New York: Viking, 2005.

Jungnickel, Christa, and Russell McCormmach. *Cavendish.* Philadelphia: American Philosophical Society, 1996.

Luckiesh, Matthew. *Artificial Light.* New York: Century Company, 1920.

Porter, George. "Joseph Priestley and His Contemporaries," *Journal of General Education* 27 no. 2 (1975): 91–100.

Robins, F. W. *The Story of the Lamp.* London: Oxford University Press, 1939.

Rossotti, Hazel. *Fire.* New York: Oxford University Press, 1993.

Sparks, Jared. *The Works of Benjamin Franklin, Vol. VI.* Boston: Hillard Gray, 1840.

Turner, Charles. *The Chemistry of Fire and Hazardous Materials.* New York: Allyn Bacon, 1981.

Walthew, Kenneth. *From Rock and Tempest: The Life of Captain George William Manby*. London: Bles, 1971.

Watts, Isaac. *The Improvement of the Mind*. Boston: James Loring, 1793.

Wilson, Mitchell. *American Science and Invention, a Pictorial History*. New York: Simon and Schuster, 1954.

Index

adhesion, 42
Ain Jalut, 132–133
air, ix, 82, 92–93, 140. *See also* hydrogen; oxygen
air pollution, 21, 156, 202
airs, doctrine of, 95, 123
Alkarsin, 161
alum, 134–135
arc lights, 123–130, 167
Argand lamps, 41
Aristotle, ix–x
atomic emissions, 175
atoms, 114

balloon aircrafts, 83
barium, 125, 137
batteries, manganese dioxide from, 97
Baybars, Sultan of Egypt, 131–133
Beddoes, Thomas, 123
Bell, Book, and Candles (rite), 48–49
Bernard, Sir Thomas, 59
Berthollet, Claude-Louis, 135
black-body radiation, 175
borax, 136
boric acid, 136, 138
boron, 125, 136–137, 138, 180
Boyle, Robert, 82
Brown, Sanborn, 56
Brush, Charles, 126
Buddhism, 48
Build a One-Candlepower Engine, 52–54

Build a Thermocouple from Scratch, 172–173
Building the Fire Tornado, 188–191
Bunsen, Robert, 157, 160–163, 176–178
Bunsen burners, 176, 177–179
Bunsen Photometer, The, 160–163
burning, vs. fire, 100, 105
Burning Ring of Fire, The, 73–78

cacodyl, 160, 161, 177
calcium, 125, 137, 180
caloric theory of heat, 61
campfires, faux, 139–145
camping stoves, 73–78
candles
    chemical equations for, 21
    experiments using, 13–20, 52–54
    Faraday lectures on, 7, 10, 11–13, 103, 156
    history of, 47–48
    incandescence and temperature, 175
    overview of chemical reactions, 20–22, 49–51, 175–176
    oxygen theory and, 92–93, 103–105
    religious rituals using, 48–49
    symbolism of, 31
    zones of burning, 50–51
capillary action, 41–42, 49, 176
carbon black, 156

carbon dioxide, 20–21, 49, 50, 100, 150, 151–153, 156, 202
Carboniferous Era, 202
carbon particles (soot), 155–156, 175–176
Carcel lamps, 41
Carl Theodore, ruler of Bavaria, 57, 58
Cavendish, Henry, x, 8, 81–85
cesium, 161, 178
charcloth, 121–122
Charles, Jacques, 83
"Chemical History of a Candle, The" (Faraday lecture), 7, 10, 12
chemicals. *See also specific names of chemicals and elements*
    flame tests and incandescent properties of, 177–178
    isolation of elements, 124–125, 136–137
    nomenclature system for, 101–102
    organometals, 160–162
Chicago fire, 184–185
chimney dampers, 70
Christmas Lectures, The, 9–13, 59–60, 103
climate change, 201–202
cloth
    combustible vs. flame-resistant types, 137
    fireproofing, 131–134, 135–137
cohesion, 41
Cold Fire, 140–145
color
    flame analysis and, 175–178
    heat and, 156
    stoplight experiments, 181–182
Constructing a Pneumatic Trough, 86–88
cracking, 50
Crookes, William, 10–11, 178
Crookes Radiometer, 10

Dalton, John, x, 8, 113–115, 116
Davy, Humphry, 85, 123–126, 136–137, 155–156, 167
dephlogisticated air, 92, 104. *See also* oxygen
Desaga, Peter, 177

Diesel, Rudolf, 114–115, 116
Disneyland attractions, 139–140

Edison, Thomas, 125
effusion, 78–79
Egyptian history, 47, 131–134
electricity, 125–126, 167–168
elements
    of chemistry, 124–125, 137, 178
    of nature, ix–x, 82, 83
Empedocles, ix
engines, thermodynamic, 54, 116
equations, 21, 105, 159
equipment, 4–5
experiments and projects
    Arc Light, 127–130
    Burning Ring of Fire, The, 73–78
    Charcloth, 121–122
    Cold Fire, 140–145
    Exploding Bubbles, 106–112
    Exploring the Interior of a Flame, 17–19
    Fire in the Hole (Making Fire from Friction), 62–67
    Fire Piston, 117–121
    Fire Tornado, 188–192
    Fireproof Cloth, 134–135
    Fire-Resistant Paper, 138–139
    Flame Test, The, 179–181
    Flame Tube, The, 22–29
    Heat Map, 19–20
    Hydrogen Generator, 88–91
    Ignite Smoke, 15–16
    manganese dioxide extraction, 97
    Mini-Extincteur, 151–154
    Olive Oil Lamp, The, 42–45
    One-Candlepower Engine, 52–54
    Oxygen Re-ignition, 99–100
    Oxygenizer, The, 96–99
    Oxyhydrogen (Exploding Bubbles), 106–112
    Parade Torch, 36–39
    Photometer Bunsen, 163–164
    Rumford, 158–160
    Pneumatic Trough, 86–88
    Propane Flamethrower, The, 193–200

Shaping a Candle Flame, 13–15
Stoplight of Fire, The, 181–182
Thermocouple building, 172–173
measuring temperature of a flame,
    169–171
Exploring the Interior of a Flame,
    17–19
extinguishers, fire, 3, 4, 147–151
Extracting Manganese Dioxide from a
    Nonalkaline Battery, 97

fabric
    combustible vs. flame-resistant
        types, 137
    fireproofing, 131, 133–135
Fahrenheit, Daniel, 166–167
Faraday, Michael
    electrical dynamo, 126
    lectures and candle experiments,
        7–13, 19, 59, 103, 156
    Manby lectures and, 150
    sound experiments, 23
    thermoelectric effect studies,
        167–168
faux fire, 139–140
festivals, Greek, 33
Feuerkoben, 115
fire, overview
    burning vs., 100
    chemistry of, 49–50, 105
    definition, 22
    descriptive language of, 183
    as element, ix–x
    history of, ix–xii, 8, 31, 39, 83
    safety, 1–6
firearms, 133
fire damp, 155
fire departments, 149
fire drills, 62–67
fire extinguishers, 3, 4, 147–151
Fire in the Hole, 612–67
fire pistons, 115–120
fireplaces, 69–72
fireproofing
    fabric, 134–135
    history of, 133–134, 135–137, 139
    paper, 138–139

fire-starting techniques, 62–67
fire tornadoes (fire whirls)
    in history, 183–186
    overview, 186–187
    project building, 188–192
fire-watching, as entertainment,
    148–149
firewood, as natural resource, 69–70
five pillars of learning, xi–xii
flame edge, 20, 21, 49, 51, 175
flame resistance
    fabric, 134–135
    history of, 133–134, 135–
        137, 139
    paper, 138–139
flames
    analysis and testing of, 177–181
    Bunsen burner, 176, 177–178
    candle, 13–15, 17–19, 21
    luminosity and, 156–164
Flame Test, The, 179–181
flamethrowers, propane, 193–200
Flame Tube, The, 22–30
Four Elements, ix–x, 83
fractious airs, 82. See also hydrogen
Franklin, Benjamin, 40–41, 55,
    70–71, 84, 102
Franklin stoves, 71
French Revolution, 101, 102–
    103, 135
friction, fire by, 62–67
fuel, 21, 79
fuel pyrolizing zones, 50–51

Galileo, 166
gases, 78–79, 123–124, 155
Gay-Lussac, Joseph Louis, 116,
    135–137, 139
global warming, 203
Goliath's Well, 132–133
Graham, Thomas, 73, 78–79
grease spot photometers, 162–163
Great Chicago Fire, 184–185
Greek history and mythology, ix,
    32–33
greenhouse effect, 21
gunpowder weapons, 57, 133–134

heat
   color and, 156
   as fire by-product, 21
   flame analysis and, 20, 21, 49–50,
      175–178
   nature of, 60–61
   pressure elasticity and, 114, 116
   temperature measurements,
      165–173
heat maps, 19–20
"History of the Corruptions of
      Christianity" (Priestly), 94
Hofmann, August Wilhelm von, 176
Home Science Tools, 5
How to Ignite Smoke, 15–16
How to Make a Parade Torch, 36–39
How to Make Charcloth, 121–122
Hulagu, 132
hydrocarbons, 21, 22, 41, 49,
      100, 105
hydrochloric acid, 83
Hydrogen Generator, 88–91
hydrogen generators, 88–91,
      108, 110
hydrogen (inflammable air, fractious
      air), 82–84, 85, 106–112
hydrogen peroxide, 137

Ilkhanate, 132
Improvement of the Mind, The
      (Watts), xi
incandescence, 175. See also color
"inflammable air," 83, 85. See also
      hydrogen
inner combustion zones, 50, 51
insects, giant, 201–202

Judaism, 48

kinetic theory of heat, 61

laboratory burners, 161, 176,
      177–178
lampas, 33
lamps
   coal miner's safety, 155
   invention of, 31–34, 39–42

laboratory burners, 161, 176,
      177–178
   olive oil lamp construction, 42–45
Lascaux, 31–32, 156
laughing gas, 124
Lavoisier, Antoine
   career, 101–103
   Cavendish and, 82, 84
   chemistry textbooks by, 135
   death of, 92, 103
   oxidation discoveries, 104–105
   personality of, 8
   Priestley's findings used by, 93
Law of Effusion, 78–79
Law of Inverse Squares, 157
Le Chatelier, Henry, 168
Leyden jars, 167
light
   arc (electric), 125–130, 167
   as electromagnetic radiation, 175
   history of, 31–32
   measuring brightness of, 158–164
   soot production and, 155, 156
luminosity, 156–160, 162–163

Making a Davy Carbon Arc Light,
      127–130
Making a Fire Piston, 117–120
Making a Heat Map, 19–20
Making a Mini-Extincteur, 151–154
Making Cloth Fireproof, 134–135
Making Fire from Friction, 62–67
Making Paper Fire-Resistant, 138–139
Mamelukes, 131–133
Manby, George, 8, 147–151
manganese dioxide extraction, 97
McMaster-Carr, 5
Measuring the Temperature of a Flame
      with a Thermocouple, 169–171
mercuric oxide, 102
methane gas, 155
metric units, 5–6
Middle Ages, x, 34, 48–49, 131–
      133, 166
molecular emissions, 175
Mongke, Great Khan of Mongol
      Empire, 132

Mongol Empire, 131–133
Montgolfier brothers, 83
mythology, Greek, ix, 32

nitrous oxide, 124

oil lamps, 33, 34, 40–45
Olive Oil Lamp, The, 42–45
"On Fractious Airs" (Cavendish), 82
organometals, 160–161
outer combustion zones, 50, 51
oxidation, 22, 50, 103–105
oxygen
    as component of air, 92, 95
    concentration in prehistoric history,
        201–202
    discovery and studies on, 92,
        104–105
    earth's present levels of, 202
    experiments and projects using,
        96–100, 106–112
    as fire component, 21, 83
    insect growth limitations and, 201
oxygen generators, 109
Oxygenizer, The, 96–99
Oxygen Re-ignition, 9-100
Oxyhydrogen, 10-112

Paleozoic Era, 202
palm boards, 67
paper, boronized, 138–139
parades, torchlight, 34–35
paraffin liquefaction zones, 50, 51
paraffin wax, 48, 49–50
phlogiston theory, x, 104–105
photometers, 158–164
Pirates of the Caribbean (Disneyland
    attraction), 139–140
Pneumatic Institution for Inhalation
    Gas Therapy, 123–124
pneumatic troughs, 85–86, 110
political campaigns, 34–35
potassium bicarbonate, 150
prehistoric history, 31–32, 39–40,
    201–202
Priestley, Joseph, x, 92–95, 104
projects. See experiments and projects

Prometheus, ix, 32–33
Promethia (Greek festival), 33
Propane Flamethrower, The, 193–200
pyrolysis, 50

Quarterly Journal of Science, The, 78
Quarterly Journal of the National
    Fire Protection Association, 138
Qutuz, Sultan of Egypt, 132–133

radiation, electromagnetic, 175
religious rites, 48–49
Rijke, Pieter, 22–23
Robert, Nicolas, 83
Roman candles, 47
Roman Catholicism, 48–49
Romans, ancient, 41, 47
Royal Institution (London)
    formation of, 59
    lecturers and researchers at,
        8, 9–10, 59, 85, 113,
        124, 150
    members (subscribers) to, 59,
        84, 85
Rubens, Heinrich, 24
Rubens Tube, 24
rubidium, 178, 180
Rumford, Benjamin Thompson,
    Count, 8, 55–62, 71–72, 84,
    157, 158
Rumford Photometer, 158–160
Rumford Roaster, 72

safety, overview, 1–6
sans culottes, 102–103
saucer lamps, 42
Science Company, The, 5
scientists
    as modern profession, 203
    perception during French
        Revolution, 103
    popularity in history, 203
    stereotypical personalities
        of, 8
    term usage, xi
Seebeck, Thomas, 165
Shaping a Candle Flame, 13–15

"Short Account of Experimental
    Researches on the Diffusion of
    Gases through Each Other, A"
    (Graham), 78
silicates, 139
silk, and cold fire, 140
smoke ignition experiments, 15–16
soda-acid extinguishers, 150–153
sodium, 178, 180
soot, 50, 155–156, 175–177
stagecraft, 139–140
Stein, Elbridge, 56
Stone Age, 31, 39
stone lamps, 40
stoplight project, 181–182
Stoplight of Fire, The, 181–182
stoves, 70–72, 73
sulfur, 100, 136, 137
supplies, 4–5

tallow, 47–48
temperature
    candle flame and, 21, 49
    color and, 175–176, 177–178
    measuring, 165–173
thallium, 176
Thenard, Louis, 135
thermocouples, 21, 167–173
thermodynamic engines, 52–54

thermometers, 166–167
Thompson, Benjamin (Count
    Rumford), 8, 55–62, 71–72, 84,
    157, 158
tinder materials, 63
torches
    construction of parade, 36–39
    history of, 32–35
    symbolism of, 32
torchlight parades, 34–35
torch-race, festival of the, 33
Traité Élémentaire de Chimie
    (Lavoisier), 135

United Nuclear, 5
units of measurement, 5–6

Volta, Alessandro, 167
voltaic pile, 167
voltmeters, 168–169

water, 20, 82, 83, 84, 114
Watts, Isaac, xi– xii, 62, 92, 115
wave form visualizers, 30
Wide Awakes, 34–35
World's Fair arc light, 126

Zeus (Greek god), ix, 32
zones of burning candle, 50–51